Silvia Hostettler

Mexican Migration

Silvia Hostettler

Mexican Migration

Reflections of Remittances in the Landscape

Südwestdeutscher Verlag für Hochschulschriften

Impressum/Imprint (nur für Deutschland/ only for Germany)
Bibliografische Information der Deutschen Nationalbibliothek: Die Deutsche Nationalbibliothek verzeichnet diese Publikation in der Deutschen Nationalbibliografie; detaillierte bibliografische Daten sind im Internet über http://dnb.d-nb.de abrufbar.
Alle in diesem Buch genannten Marken und Produktnamen unterliegen warenzeichen-, marken- oder patentrechtlichem Schutz bzw. sind Warenzeichen oder eingetragene Warenzeichen der jeweiligen Inhaber. Die Wiedergabe von Marken, Produktnamen, Gebrauchsnamen, Handelsnamen, Warenbezeichnungen u.s.w. in diesem Werk berechtigt auch ohne besondere Kennzeichnung nicht zu der Annahme, dass solche Namen im Sinne der Warenzeichen- und Markenschutzgesetzgebung als frei zu betrachten wären und daher von jedermann benutzt werden dürften.

Verlag: Südwestdeutscher Verlag für Hochschulschriften Aktiengesellschaft & Co. KG
Dudweiler Landstr. 99, 66123 Saarbrücken, Deutschland
Telefon +49 681 37 20 271-1, Telefax +49 681 37 20 271-0, Email: info@svh-verlag.de
Zugl.: Lausanne, Ecole Polytechnique Fédérale de Lausanne, Thèse Nr 3730, 2007

Herstellung in Deutschland:
Schaltungsdienst Lange o.H.G., Berlin
Books on Demand GmbH, Norderstedt
Reha GmbH, Saarbrücken
Amazon Distribution GmbH, Leipzig
ISBN: 978-3-8381-0443-0

Imprint (only for USA, GB)
Bibliographic information published by the Deutsche Nationalbibliothek: The Deutsche Nationalbibliothek lists this publication in the Deutsche Nationalbibliografie; detailed bibliographic data are available in the Internet at http://dnb.d-nb.de.
Any brand names and product names mentioned in this book are subject to trademark, brand or patent protection and are trademarks or registered trademarks of their respective holders. The use of brand names, product names, common names, trade names, product descriptions etc. even without a particular marking in this works is in no way to be construed to mean that such names may be regarded as unrestricted in respect of trademark and brand protection legislation and could thus be used by anyone.

Publisher:
Südwestdeutscher Verlag für Hochschulschriften Aktiengesellschaft & Co. KG
Dudweiler Landstr. 99, 66123 Saarbrücken, Germany
Phone +49 681 37 20 271-1, Fax +49 681 37 20 271-0, Email: info@svh-verlag.de

Copyright © 2009 by the author and Südwestdeutscher Verlag für Hochschulschriften Aktiengesellschaft & Co. KG and licensors
All rights reserved. Saarbrücken 2009

Printed in the U.S.A.
Printed in the U.K. by (see last page)
ISBN: 978-3-8381-0443-0

Abstract

The present study focuses on the economic, political/institutional, technological, cultural, demographic and environmental drivers of land use change. It aims to understand the factors influencing land use decisions at the household level, in particular the influence of migration. The study is guided by the hypothesis that international migration is driving land use change through the investment of remittances, funds sent back by migrants to their families in the country of origin. This research is based on a political ecology approach and the conceptual framework relies on three theoretical concepts. First, the concepts of *proximate causes* and *driving forces* were used to identify the factors behind changing land use. In addition, the concept of *remittance landscapes*, a concept developed in the framework of this study, which is defined as *an emerging type of landscape driven by the investment of remittances*, was used to evaluate the impact of remittances on land use in the study area.

Fieldwork was conducted in the municipality of Autlán in the state of Jalisco in Mexico over a total period of 8 months between 2002 and 2004. Land use changes between 1990 and 2000 were quantified based on satellite image analysis. Underlying driving forces of these changes were examined based on land use change data collected by survey as well as data available from municipal, state and federal agencies.

Land use changes observed in the study area between 1990 and 2000 include a slight increase of agricultural land (2%), of urban land cover (0.5%) and of pine-oak forest (0.7%). Over the same period, pasture increased by 18% while dry forest decreased by 10%. Rapid and extensive land use change is occurring on rainfed agricultural land, as maize cultivation is converted to the cultivation of *agave azul* used for the production of tequila. The first plantations of *agave azul* were established in 1996 and by 2002, *agave azul* was planted on 33% of all rainfed agricultural land of the municipality. 84% of owners of rainfed land included in the survey had changed land use from maize to agave during this time period.

The dynamics of several proximate causes are driving this change: 1) Market prices for maize decreased by 46% between 1994 and 2004 while the costs for agricultural inputs continually increased so that the cultivation of rainfed maize was no longer economically profitable; 2) The variability of rainfall combined with a lack of irrigation water limits the choice of economically viable alternatives to *agave azul*; 3) In the large majority of cases, landowners rent out their land to tequila companies in reverse leasing arrangements for seven-year periods (the duration of one growing cycle of *agave azul*). During this time they do not have to work on their own fields and are free to find off-farm employment or to migrate to the US and; 4) Landowners continue to receive agricultural subsidies even though the land is rented out, as *agave azul* is one of the eligible crops.

Overall, the main driving forces identified in the study area are economic (market prices), environmental (variability of rainfall, soil quality, topography), political/institutional (agricultural subsidies, land tenure) and demographic (labor availability). Technology and culture appear to be less important. Results of the present study confirm the hypothesis that global factors, especially international trade agreements such as NAFTA (North American Free Trade Agreement) increasingly influence land use change. However, they are not sufficient to function as a sole driver of land use change. Environmental factors are a critical determinant of whether a certain land use change will occur or not. The decisive aspect behind the observed land use changes are the multiple interactions between specific factors at different levels and not the predominance of one particular driving force functioning at a particular level.

International migration is a significant livelihood strategy in the study area, especially for lower-income communities. On average, 50% of all households have or had at least one family member in the US as a migrant between 1980 and 2004, and remittances represent 45% of total household income. In general, the bulk of remittances income is used for subsistence needs and to repay debts. Nevertheless, on average, 30% of migrant households invest remittances in land, livestock, agricultural production and in house construction. All these investments lead to land use changes. The impact of remittances on land use changes is variable, and depends on the socio-economic, political and environmental context of the community and the individual situation of the migrant household. In low-income communities, remittances might be used to repair existing housing, while in higher-income communities, remittances are used to construct a new house, converting agricultural to urban land. With regard to changes in labor availability due to out-migration, the results are ambiguous. Migration can drive land use change by encouraging a shift to low-labor land use systems, but these land use changes that require less labor can also drive migration.

The concept of remittance landscape developed by the researcher has proved useful for analysing the impact of remittances on land use changes. A combination of area-based and actor-based evaluation criteria are effective in order to describe quantitative as well as qualitative landscape transformations driven by the investment of remittances. Landscapes where the investment of remittances leads to a change of land use from subsistence to cash crop cultivation should be included as a potential type of remittance landscape, even though the basic type of the landscape (agricultural) remains unchanged.

Key words: Land use change, driving forces, migration, remittance landscape, Mexico

Résumé

Cette étude se concentre sur les facteurs économiques, politiques/institutionnels, technologiques, culturels, démographiques et environnementaux qui génèrent un changement d'utilisation du sol. Elle a pour objectif de comprendre les facteurs qui influencent l'utilisation du sol au niveau du ménage, et plus particulièrement celui de la migration. L'étude est guidée par l'hypothèse que la migration internationale est une force motrice du changement d'utilisation du sol par l'apport de fonds étrangers envoyés par les migrants à leurs familles restées dans leur pays d'origine (*remittances*). Cette recherche est basée sur une approche d'écologie politique et son cadre conceptuel repose sur trois concepts théoriques. Les concepts de *causes immédiates* et *forces motrices* sont utilisés pour identifier les facteurs sous-jacents du changement d'utilisation du sol. De plus, le concept de '*remittance landscape*'- défini comme '*un nouveau type de paysage dû à l'impact de l'investissement de fonds étrangers envoyés par les migrants*' a été développé dans le cadre de cette étude et utilisé pour évaluer l'impact de ces fonds sur l'utilisation du sol dans la région de l'étude.

Entre 2002 et 2004, huit mois de travail sur le terrain ont été conduits dans la municipalité d'Autlán dans l'état de Jalisco en Mexique. Les changements d'utilisation du sol entre 1990 et 2000 ont été quantifiés à l'aide de l'analyse d'images satellites. Les changements d'utilisation du sol observés dans la région d'étude entre 1990 et 2000 indiquent une légère augmentation des terres agricoles (2%) et des surfaces urbaines (0.5%) et des forêts de pins-chênes. Pendant la même période, les pâturages ont augmenté de 18% alors que la forêt tropicale sèche diminuait de 10%. Un changement rapide et extensif est en train de se produire sur les terres agricoles non irriguées, car la culture de maïs est convertie en culture d'*agave azul*, utilisé pour la production de tequila. Les premières plantations d'agave *azul* datent de 1996 et en 2002, l'*agave azul* est planté sur 33% de toute les terres agricoles non-irriguées de la municipalité. 84% des propriétaires des terres non irriguées inclus dans l'enquête ont changé d'utilisation du sol et sont passés de la culture du maïs à celle de l'agave sur cette période. Les dynamiques entre les causes immédiates suivantes génèrent ce changement: 1) Le prix du maïs a diminué de 46% entre 1994 et 2004 alors que les coûts des facteurs de production agricole ont augmenté continuellement de telle sorte que la culture du maïs sur les terres non-irriguées n'est plus rentable; 2) L'irrégularité des précipitations et l'absence d'irrigation posent de réelles limites quant aux alternatives économiquement viables à la culture de l'*agave azul*; 3) En large majorité, les propriétaires fonciers louent leurs terres aux entreprises de tequila pour des périodes de sept ans qui correspondent à un cycle de maturation de l'agave. Pendant ce temps, ils ne travaillent pas leurs propres terres et sont libres de trouver du travail à l'extérieur de leur ferme ou de migrer aux Etats-Unis et; 4) Comme l'*agave azul* est une culture éligible pour les subventions agricoles, les propriétaires fonciers continuent de recevoir les subventions malgré le fait que leurs terres soient louées.

En général, les forces motrices majeures identifiées dans la région d'étude sont économiques (prix du marché), environnementales (l'irrégularité des précipitations, la qualité du sol, la topographie), politiques/institutionnelles (subventions agricoles, régime foncier) et démographique (force de travail disponible). Les facteurs technologiques et culturels paraissent moins importants. Les résultats de cette étude confirment l'hypothèse que l'influence des facteurs globaux augmente dans les changements d'utilisation du sol, et plus particulièrement celle des accords de commerce internationaux comme le NAFTA (Accord de libre échange nord-américain). Pourtant, les facteurs globaux ou économiques ne sont pas suffisants pour agir comme seules forces motrices du changement. L'aspect décisif de cette étude sur les changements d'utilisation du sol est de montrer que les interactions sont multiples entre des facteurs spécifiques, localisés à plusieurs niveaux et qu'il n'y a pas de prédominance d'une seule force motrice agissant à un niveau particulier.

La migration internationale est une importante stratégie de vie dans la région d'étude, en particulier dans les communautés à bas revenus. En moyenne, 50% des ménages ont ou ont eu au moins un membre de leur famille émigré aux Etats-Unis entre 1980 et 2004, et les envois de fonds représentent 45% du revenu total du ménage. En général, la majorité des envois de fonds est utilisée à des fins de subsistance et de remboursement de dettes. Néanmoins, en moyenne, 30% des familles des migrants investissent les apports financiers de l'étranger pour acheter des terres, du bétail, des investissements agricoles et pour la construction d'une maison. Tous ces investissements entraînent un changement d'utilisation du sol. L'impact des envois de fonds sur l'utilisation du sol est variable et dépend largement du contexte socio-économique, politique et environnemental de la communauté ainsi que de la situation individuelle du ménage. Dans les communautés à bas revenu, les fonds envoyés sont utilisés, par exemple, pour réparer une maison existante alors que dans les communautés à haut revenu, les envois sont utilisés pour construire une nouvelle maison, ce qui implique une conversion de la terre agricole en terre urbaine. Par rapport aux changements dans la disponibilité de la force de travail du à l'émigration, les résultats sont ambigus; l'émigration peut inciter au changement d'utilisation du sol en encourageant un changement vers un système nécessitant peu de travail, mais ces même systèmes qui ont besoin de peu de travail peuvent aussi inciter à l'émigration.

Le concept de '*remittance landscape*' développé par la chercheuse a démontré son utilité pour analyser les impacts des fonds envoyés par les migrants sur le changement d'utilisation du sol. Les transformations de paysage vers un '*remittance landscape*' sont efficacement décrites par une combinaison de critères d'évaluation basée sur la surface affectée par le changement et la proportion d'une population prenant part à une pratique agricole qui transforme le paysage. Les paysages dans lesquels l'apport de fonds provenant de l'étranger provoque un basculement d'une agriculture de subsistance vers une agriculture commerciale doivent être inclus en tant que type potentiel de '*remittance landscape*' malgré le fait que le type fondamental du paysage (agricole) reste inchangé.

Mots-clés: Changement d'utilisation du sol, forces motrices, migration, remittance landscape, Mexique

Acknowledgements

The work presented here would not have been possible without the financial, logistical and intellectual support of many others. Funding was provided by the Swiss National Centre of Competence in Research North-South (NCCR North-South): Research Partnerships for Mitigating Syndromes of Global Change. The NCCR North-South programme is co-funded by the Swiss National Science Foundation and the Swiss Agency for Development and Cooperation (SDC). The Ecole Polytechnique Fédérale de Lausanne (EPFL) has also funded part of this study which is duly acknowledged. I am very grateful to my advisor Prof. Jean-Claude Bolay and co-advisor Prof. Hans Hurni who were very supportive through all the various permutations of this research, from inception to analysis, and their journey to my field site in Mexico was thoroughly enjoyed and appreciated.

I am deeply indebted to Claudia Pelayo Guzmán, who was an invaluable field assistant, very resourceful finding the right people and procuring useful data, patient with the many interviews and a great friend throughout this study. Special thanks to her large family and in particular to Doña María Vidriales who, despite her 80 years, walks up mountains quicker than most and without losing her breath either.

The people of El Jalocote, Chiquihuitlán and Mezquitán have my gratitude for their hospitality, patience, and curiosity; having the opportunity to meet them and hear their stories was the most fascinating part of this research.

I would like to thank Dr. Peter Gerritsen and Claudia Ortíz from the Manantlán Institute of Ecology and Conservation of Biodiversity (IMECBIO) of the University of Guadalajara. They provided valuable guidance, organizational and logistical support. I would also like to thank Dr. Oscar Cárdenas of the University of Guadalajara for his help with satellite imagery. I am also indebted to Luis Eugenio Rivera, Luis Manuel Martínez, Arturo Solís, Salvador García, Martin Gómez Garcia and Sergio Graf of IMECBIO as well as Juan Ignacio Arroyo, Adrian Murguía and José Luis Dominguez from the municipality of Autlán for their time and endurance answering my numerous questions. Also in Mexico, I would like to thank Barbara Gröhling for being an inspiring friend and for providing a peaceful environment by the sea during the weekends.

Abram Pointet from LASIG (EPFL) provided expertise for satellite image classification and Dr. Andreas Heinimann (CDE, University of Berne) produced the Digital Terrain Model and land use change maps. Many thanks to both for their assistance and patience with the endless details of spatial analysis. Thanks also to Anna Svanberg who spent many hours digitizing the satellite images.

I would like to thank Brigitte Portner for providing the data for chapter 5 of this thesis and for reading and commenting on draft chapters of this dissertation. Thanks to Dr. Christine Bichsel and Dr. Balz Strasser for friendship, advice and intellectual support.

Thanks to Dr. Laura Ediger for interesting discussions on conceptual frameworks, statistical illumination and above all for proofreading of this thesis.

I would also like to acknowledge Jonathan Ball, Sophie Chaubaroux, Julie Gaudreau, Dr. Tobias Hagmann, Prof. V. Kaufmann, Dr. Eva Ludi, Roberta Méan, Dr. Peter Messerli, Prof. Ulrike Müller-Böker, Dr. Yves Pedrazzini, Dr. Stephan Rist, Dr. Tina Svan Hanson, Dr. Susan Thieme, Dr. Bettina Wolfgramm, Michelle Young and Dr. Anne Zimmermann who provided support in one way or another during this research.

Thanks to family and friends, especially to my sisters Barbara and Stefanie for their continued support and occasional distractions. Thanks to Patrick for his assistance with graphics and layout, which is greatly appreciated. Finally, I am entirely indebted to our children Noé and Ella who are a constant source of joy and invaluable for keeping things in perspective.

Contents

Abstract	i
Résumé	iii
Acknowledgements	v
Contents	vii
List of tables	ix
List of figures	xi

PART I   1

1. **Introduction**   3
    1.1 *Driving forces of land use and cover change*   4
    1.2 *Mexican migration to the United States*   5
    1.3 *Study background*   6

2. **Study area**   9
    2.1 *Land tenure system in Mexico*   9
    2.2 *The municipality of Autlán*   11
    2.3 *Case study sites*   13

3. **Theoretical approach**   21
    3.1 *The Land Use and Land Cover (LUCC) project*   21
    3.2 *The impact of remittances*   22
    3.3 *Investment patterns of remittances*   25
    3.4 *Factors determining investment patterns of remittances*   28
    3.5 *Migration concepts and definitions*   30
    3.6 *Theory of land use change*   32
    3.7 *Political ecology*   34
    3.8 *Research questions and hypothesis*   36

4. **Conceptual framework**   39
    4.1 *The concept of proximate causes and driving forces*   39
    4.2 *The concept of remittances landscapes*   41
    4.3 *Methodological framework*   42
        4.3.1 *Spatial assessment of land use changes*   44
        4.3.2 *Interview-based assessment of land use changes*   47

PART II   51

5. **Differences in land use strategies between migrant and non-migrant households**   53
    5.1 *Livelihood capitals*   55
    5.2 *Case examples*   56
        5.2.1 *Innovative migrants*   56
        5.2.2 *Non-innovative migrants*   60
    5.3 *Conclusions*   64

6. Land use changes 1980-2002 — 65
6.1 Land use changes at municipal level 1990-2000 — *65*
6.2 Land use changes to agave azul 1996-2002 — *71*
6.3 Land use changes in agricultural communities and on private properties — *73*
    6.3.1 Land use changes on land plots (parcelas) — *74*
    6.3.2 Land use changes in the hills (cerro) — *75*
6.4 Assessment of effects of land use change to agave azul — *76*
6.5 Conclusions — *81*

7. Proximate causes and dynamics of driving forces influencing land use change — 83
7.1 Methods of analysis — *83*
7.2 Case examples from El Jalocote — *84*
7.3 Case examples from Chiquihuitlán — *96*
7.4 Case examples from Mezquitán — *102*
7.5 Case examples of private property owners — *107*
7.6. Dynamics of proximate causes and driving forces — *113*
7.7 Conclusions — *118*

8. The impact of migration: Remittances as drivers of land use change — 121
8.1 Characteristic of transnational migration from Mexico to the US — *121*
8.2 Impacts of transnational migration on land use changes — *127*
8.3 Investment of remittances — *128*
8.4 Impact of remittances on land use — *132*
8.5 Conclusions — *138*

PART III — 141
9. Synthesis — 143
9.1 Driving forces at the global level — *143*
9.2 Driving forces at the national level — *144*
9.3 Future trends in land use — *146*
9.4 The impact of migration and remittances — *148*
9.5 Recommendations — *152*
9.6 Further research needs — *153*

10. Conclusions — 155

References — 161
Glossary / definitions — 177

Annex — 181
Annex 1 Cross-tabulation matrix of land use changes 1990-2000 — *183*

List of tables

Table 4-01	Description of land use classes used for classification of satellite images	44
Table 4-02	Overview of interviews conducted	49
Table 4-03	Overview of proximate causes of driving forces	50
Table 6-01	Land use change between 1990 and 2000 in the municipality of Autlán	68
Table 6-02	Comparison of land use in 1993 and 2000 at national level in Mexico	68
Table 6-03	Percentage of landowners changing land use, 1980-2002	74
Table 6-04	Percentage of households (HH) having changed land use on agricultural land plots between 1980-2000	75
Table 6-05	Percentage of households (HH) having deforested land in the hills between 1980-2002	76
Table 6-06	Description of various degrees of soil erosion	80
Table 6-07	Percentage of fields per slope class and degree of erosion	80
Table 7-01	Family tree of Doña Mariana Morales	87
Table 8-01	Occupation of migrants in the US (in %)	123
Table 8-02	Reasons for migration (in %)	124
Table 8-03	Migration characteristics	124
Table 8-04	Reasons for not migrating (in %)	125
Table 8-05	Changes in work load for members of migrant household (in %)	126
Table 8-06	Differences in migration activity between landless and landowning households (in %)	128
Table 8-07	Significance of remittances as source of income (in %)	129
Table 8-08	Impact of remittances on situation of household	130
Table 8-09	Impact of remittances on land use changes	133
Table 8-10	Main differences in context and investment patterns of remittances between case study sites	137

List of figures

Figure 2-01	Map of Mexico	9
Figure 2-02	Map of the state of Jalisco	13
Figure 2-03	Municipality of Autlán (State of Jalisco, western Mexico)	15
Figure 4-01	Visualization of conceptual framework	41
Figure 4-02	Coverage of contour lines	46
Figure 5-01	Study locations of La Laja de Abajo and Rincón de Luisa	53
Figure 5-02	Capitals of Carlos García	58
Figure 5-03	Capitals of Gustavo Velázquez	60
Figure 5-04	Capitals of Isidro Santana	62
Figure 5-05	Capitals of Luz Parrado	63
Figure 6-01	Land use in the municipality of Autlán 1990 and 2000	67
Figure 6-02	Number of cattle in the municipality of Autlán and the state of Jalisco (1993-2002)	70
Figure 6-03	Area cultivated with various crops and pasture in the municipality of Autlán (1990 - 2002)	71
Figure 6-04	Location of agave fields in municipality	78
Figure 6-05	Distribution of fields (n = 224) per slope class	79
Figure 7-01	Average, maximum and minimum monthly rainfall in Mexico (1941-2001)	114
Figure 7-02	Rainfall anomaly March-May 1941 to 2004 (Region VIII)	115
Figure 7-03	Proximate causes and driving forces of land use change in the hills	117
Figure 7-04	Proximate causes and driving forces of land use change on agricultural land	118
Figure 8-01	Investment of remittances	131
Figure 10-01	Overview of proximate causes and driving forces in study area	156

PART I

1. Introduction

1.1 Driving forces of land use and cover change

Land use and land cover change emerged as an important topic on the international research agenda in the 1970s, as a consequence of the concern for the impact land use and land cover changes might have on regional and global climate (Houghton et al. 1985, Lambin et al. 2003, Woodwell et al. 1983). Land use and land cover change is an extensive and accelerating process, in many cases negatively affecting natural resources such as soil and water resources. It is driven by human actions and often also drives changes that impact humans. Investigating the driving forces of these changes is critical for formulating effective policies and for identifying the factors that encourage or impede their implementation. The impact of land cover changes is important for society. For instance, the conversion of forested areas into other uses contributes to climate change and to a loss of biological diversity (Lepers et al. 2005, WRI 2000). By altering ecosystem functions, changes in land use and land cover affect the ability of ecological systems to support human needs and such changes also determine, in part, the vulnerability of places and people to climatic, economic, or socio-political perturbations. For example, biodiversity loss due to deforestation results in a decline in ecosystem integrity and may impact hydrological processes, leading to flooding and soil erosion (FAO 2000b, Houghton 1994, Millennium Ecosystem Assessment 2005, UNEP 1999, Vitousek et al. 1997). The need to understand the driving forces behind land use changes becomes more imperative as those changes become more rapid (Johnson et al. 1999, Lambin 1993, Stromph et al. 1994). Substantive research on land use and land cover changes was requested by the international scientific community during the 1972 Stockholm Conference on the Human Environment and again in 1992 at the UN Conference on Environment and Development.

Land use and cover change is driven by a variety of socioeconomic, political, cultural, technological and biophysical factors (Agarwal et al. 2002, Bürgi et al. 2004, Lambin et al. 2003). The relative importance of a single factor or a combination of factors appears to vary widely between studies. For instance, drivers can include physical attributes of the landscape such as topography (Pan et al. 1990, Silbernagel et al. 1997), land tenure (Southworth and Tucker 2001, Turner et al. 1996), institutions and social services (Serneels and Lambin 2001), water availability (Fox et al. 2003), the general socioeconomic situation of households (Moran et al. 2003, Walker et al. 2002), or economic opportunities mitigated by institutions (Lambin et al. 2001). The relative importance of different drivers depends very much on the scale of analysis (Gibson et al. 2000, Millington et al. 2003, Walsh et al. 1999) and on the socio-ecological context (Evans et al. 2003, Pan and Bilsborrow

2005). For instance, driving forces identified at the national level may be different from those at the local level, just as the driving forces identified in a tropical forestry system (Geist and Lambin 2002) are different from those in a temperate agricultural context (Bürgi and Turner 2002). Whereas the relative importance of different drivers is often debated, there is widespread agreement that explanatory variables for a given phenomenon change as the scale of analysis changes (Gibson et al. 2005). At the same time, driving forces may differ, even at a particular scale of analysis. As an example, certain land use change studies at watershed level identified policy as the dominant driver (Gautam et al. 2002) while in another watershed, population growth was the most important factor (Bewket 2001). Furthermore, there is growing evidence that land use changes are increasingly linked to globalization (Lambin et al. 2001, Verbist et al. 2005). Developments in one part of the world affect another, and globalization increasingly separates places of consumption from places of production. Therefore, land use change is often linked to processes operating globally, and research is needed to understand how these global processes influence land use practices, including their impacts on ecosystem services (Barrett et al. 2001, Bebbington and Batterbury 2001, Bruinsma 2003, Global Land Project 2005, Mertz et al. 2005).

International migration can influence land use in the sending communities in a number of ways. Even though migration may potentially trigger land use change by increasing household income (Stéphenne and Lambin 2001, Verburg et al. 1999), it is usually overlooked in land use change studies (de Haan 1999, Lopez et al. 2006). In some rural areas, population has decreased due to out-migration. This has led to a partial abandonment of poor quality rainfed agricultural land and an expansion of scrubland, improving the overall environmental conditions (López et al. 2006). A similar trend has been observed in China (Ediger 2006) where high rates of out-migration have led to the conversion of agricultural land to forest. High rates of out-migration are often caused by a strong desire of villagers to leave agriculture (de Haan 1999, Ediger and Huafang 2006). In the present study, the potential impacts of remittances on land use will be examined, namely, the influence of migration in providing the household with an additional source of income that may provoke land use changes.

1.2 Mexican migration to the United States

The first large migration movement from Mexico to the United States began around 1900. The push factor behind this movement was primarily the Mexican Revolution. The pull factor was the demand for Mexican labor to work on the extension of the railroads and in the agricultural sector due to labor shortages in the United States as a result of World War I (Verduzco and Unger 1998).

In the 1930s, the economic depression in the US caused a population countermovement and 47% of all Mexicans who had obtained legal status since 1900 moved back to Mexico (Verduzco and Unger 1998). As the United States entered World War II, demand for Mexican laborers increased again, and prompted the establishment of the *Bracero* program. Between 1942 and 1946, 4.6 million Mexicans were contracted to work legally in the US agricultural sector under this program (García y Griego 1998). Even though they were expected to be temporary residents in the US, many stayed along with their families. The *Bracero* program functioned through migration stations in Mexico, where rural farm workers were selected as *bracero* candidates. They were then transported to reception centers at the border, where representatives of American growers hired them for relatively short periods, often not more than six weeks. The US government represented the interests of the farm employers, while Mexican officials represented *bracero* interests and negotiated employment and living conditions. Contracts for individual workers were sometimes renewed and extended. Otherwise, workers were returned to Mexico at the cost of the employer at the end of their contract (García y Griego 1998). The *Bracero* program was accompanied by a large flow of unauthorized workers from Mexico who sometimes worked for growers employing *braceros*, and some of these were awarded legal status after 1945. Although American growers put pressure on Congress to continue the supply of cheap Mexican labor, Congress refused to extend the *Bracero* program beyond 1964 (Bean et al. 1998). Despite the end of this program, both legal and illegal migration from Mexico has continued to grow. In particular, the size of the legal population increased dramatically during the late 1980s, in part due to the Immigration Reform and Control Act (IRC) under which 2.24 million Mexicans were legalized in 1986 (Verduzco and Unger 1998).

Migration to the US continues to increase (CONAPO 2003), but migration patterns and the impacts of migration in Mexico are changing. Mexico experienced sustained growth in the 1940s, during the same years the *Bracero* program was launched. This period of the "Mexican miracle" led to improvements in health care, education and better living conditions despite high rates of population growth. During this time, remittances were also important to support Mexican modernization but were not a crucial factor in the rapid growth of the Mexican economy (Verduzco and Unger 1998). Since the 1990s, when the economic crisis in Mexico began, the significance of remittances as an economic resource has continued to gain in importance, at the household as well as the national level. Between 1965 and 1985, international migration mainly consisted of small-scale rural farmers engaging in temporary migration to the US (Massey 1985, Massey et al. 2002). From 1995 onwards, migration has become much more complex: migrant sending areas have increased, urban migration has increased, the proportion of female and indigenous migrants has increased, the average stay of migrants has become longer and their legal status more fragile (Durand and Massey

2003, Lopez et al. 2006). The main reasons for migration are an absence of jobs, low salaries, lack of health benefits and above all inadequate government policies in Mexico (Massey et al. 2002). The state of Jalisco in west-central Mexico has an emigration rate of 65%,[1] similar to other states in this region (Aguascalientes (73%), Durango (59%), Guanajuato (59%), Michoacán (63%), Zacatecas (75%) (López et al. 2006).

According to the Immigration and Naturalization Department of the United States, 88% of male and 73% of female Mexicans entering the US have no papers. Each day around 400 illegal Mexican migrants are apprehended at the border trying to cross into the US. This means that out of 600 Mexicans trying to cross illegally, 200 make it while the remaining 400 are caught (The Economist 2002). In 2002, an undocumented Mexican worker in the United States was estimated to send back between USD 200 (Garcia-Zamora 2006), USD 385 (Orozco 2002) and USD 500 (IIED 2003) per month. A number of community studies show that a large part of remittances are spent on family maintenance and health; the purchase, construction and improvement of homes; and the purchase of consumer goods (Durand and Massey 1992). Migrants in their 5th to 10th year abroad send the most. Recent migrants and migrants who stay for more than 10 years send below-average amounts (Orozco 2003).

1.3. Study background

This research project was conducted under the framework of the participation of the Laboratory of Urban Sociology (*LASUR*) of the Ecole Polytechnique Fédérale de Lausanne (EPFL) in the National Centre of Competence in Research North-South (NCCR North-South): Research partnerships for mitigating syndromes of global change. The NCCR North-South research programme is funded by the Swiss National Science Foundation, the Swiss Agency for Development and Cooperation and the participating organizations.

The NCCR North-South programme focuses on research in the field of development studies. It understands negative aspects of 'global change' as problems of unsustainable development. During the first phase of the NCCR North-South (2001-2005), a list of 30 core problems was established (Messerli and Wiesmann 2004). These problems are perceived to be closely related and appear in similar combinations or clusters in specific socio-ecological contexts. Such a cluster of core

---

[1] In this context, the emigration rate of 65% means that in 65% of all the municipalities in the state of Jalisco, some of the population are active in international migration. Correspondingly, this means that in the remaining 35% of the municipalities in Jalisco, there is no international migration activity (López et al. 2006).

problems is designated as a 'syndrome of global change' (Petschel-Held et al. 1995, WBGU 1997). The NCCR North-South goes beyond analysis of core problems by aiming to develop syndrome mitigation strategies, thus attempting to reduce the negative effects of single or clustered problems (Hurni et al. 2004).

Contribution to the NCCR North-South programme

This study contributed to the NCCR North-South in two ways – first, by contributing to the research of two Institutional Partners of the NCCR North-South and second, by conducting an additional research project that partly overlapped with the present research project, applying the conceptual framework of the NCCR North-South.

Research undertaken under the framework of this PhD contributed to a research project of the Institutional Partner *LASUR* (IP5), "Integrated analysis of urbanization processes on natural resource management: The case of the lower Ayuquila river watershed in Western Mexico" (Bolay et al. 2004). The main objective of this project was to examine rural-urban interactions, such as migration, in order to evaluate their effects on natural resource management. This project was developed by *LASUR* in partnership with the Manantlán Institute of Ecology and Conservation of Biodiversity (IMECBIO) of the University of Guadalajara in Mexico. The research project was interdisciplinary, bringing together researchers from urban and rural sociology, environmental science, geography, economics, ecology, and hydrology.

This research project is also linked to the Institutional Partner *Centre for Environment and Development* (IP2) of the University of Berne. The objectives of the IP2 are to enhance knowledge on sustainable land management options for natural resources such as soil, water, vegetation and fauna in their ecological settings, thereby minimizing problems of land degradation, loss of biodiversity, and mismanagement of water (SARPI 2000).

The present study contributes to objectives (c)[2] and (e)[3] of IP2 by quantifying land use changes and investigating the underlying driving forces.

---

[2] „Adopt, adapt, and develop useful models to improve process-based understanding, and to generate knowledge of spatial units and temporal scales where little or no data exist" (SARPI 2000).

[3] „Develop and test methodologies and tools both generic and adaptive, in different disciplinary (e.g. erosion assessment and modelling methods, methods to experiment with sustainable land management, or tools to assess land use change), as well as interdisciplinary fields (e.g. geo-referenced data storage, retrieval, and use for defined spatial units)" (SARPI 2000).

Global change research focuses on human-nature interactions (Reusswig 1998). In the present research project, land use changes are analyzed as the centerpiece of human-nature interactions. This project analyzes international migration, which was designated as a core problem of sustainable development by the NCCR North-South (Hurni et al. 2004). In addition, it gives an indication as to which of the forces driving land use change are linked to global change.

From 2003-2005, together with two other NCCR North-South PhD researchers (Dr. Christine Bichsel from SwissPeace Foundation and Dr. Balz Strasser from the University of Zürich), a joint research project on labor migration in Mexico, India and Kyrgyzstan was conducted. This project contributed to the development of the NCCR North-South programme by testing the main hypothesis of the syndrome mitigation concept for the case of international labor migration. Also, the paper contributed to the discussion of an NCCR North-South methodological approach by applying a specific comparative research design. The study was published in the NCCR North-South dialogue series (Bichsel, Hostettler and Strasser 2005).

2. Study Area

Figure 2-01: Map of Mexico

Source: www.pickatrail.com

2.1 Land tenure system in Mexico

The land tenure system in Mexico is very detailed and requires a brief explanation. The Mexican Constitution of 1917 recognizes three forms of rural property: private, *ejido* and agrarian communities, which have collective ownership of land and resources (Appendini 2002).

Forms of rural property in Mexico.
Small private property (pequeña propiedad rural). This is private property and the owner has, in addition to the right of use and usufruct, the right of sale or disposal. The Constitution limits the amount of land allowed in private holdings to 100 hectares of irrigated or very humid land, with the exception of specific crops. A hectare of irrigated land equals two of rainfed land, four hectares of good-quality pasture and eight hectares of marginal or arid lands. Land for cattle-raising is limited to the amount required to graze 500 head of cattle or the equivalent amount of smaller species according to a grazing coefficient set by region. Forest property is limited to 800 hectares. Rural private property is registered at the private land registration agency.

> *Ejido.* This type of community was created by land distribution under agrarian reform (1917-1992). Land was given to the members of an *ejido* for use and usufruct, but remains the property of the nation, although the rights are inheritable. Hence *ejido* property rights are limited. Under the 1992 Agrarian Law, the Assembly of *ejido* members can decide by majority of vote to change the tenure regime.
>
> *Comunidad agraria.* Agrarian communities are collective owners of their land under a common property regime, with titles bestowed by the Spanish Crown during the Colonial period. Some *comunidades* have remained intact through the centuries, but the majority lost the titles to their land over time. Restitution of land is the mechanism by which Agrarian Reform restored the *comunidades* access to land.

Source: Appendini 2002

In Mexico, 50% of the arable land is owned by the so-called 'social sector,' consisting of rural communities organized into *ejidos* and *comunidades agrarias* (Young 2002). In 1992, the agrarian legislation was reformed to extend the limited property rights of rural land, which had been distributed to peasants over a period of 75 years since the Mexican Revolution in 1910. Prior to 1910, approximately 260 families owned 80% of the Mexican territory (Young 2002). Under the new Agrarian Law, peasants retain full property rights over their plots of land and the right to decide the future use of common lands and resources. This reform was driven by the desire to create an active land market, promote efficient resource allocation and to encourage investment in agriculture (Appendini 2002).

Land tenure in Mexico is linked to a very complex institutional context and plagued by unresolved rights and legal disputes. In 1992, when land distribution came to an end, the land tenure regularization program PROCEDE[4] was launched in order to define clear property rights by providing proper land titles. The Procuraduría Agraria (Agrarian Attorney) was established and designated responsible for settling disputes over rights and boundaries related to land tenure in *ejidos* and *comunidades agrarias*. With the end of land distribution, security of land tenure became a priority and PROCEDE is the major regularization program still active today. The objectives of PROCEDE are to give the *ejidatarios* land tenure security through legal recognition of *ejidal* property types, including the rights to individual plots, to common lands and to the titling of urban house plots. PROCEDE not only confirms property rights by establishing ownership certificates, but also aims to reinforce the structures and internal organization of the *ejidos* with regard to decision-making and conflict resolution. In this way, it was hoped that *ejidos* would become less dependent on public institutions and rely more on local participation and democracy. The program

---

[4] Programa de Certificación de Derechos Ejidales y Titulación de Solares.

aims to regulate the land tenure situation of agricultural plots and urban plots and to trace the boundaries of common lands in all 27,252 *ejidos* and 2,194 *comunidades* in Mexico, which constitutes an enormous task. Since private properties are in the Public Property Register, they are not part of this program.

PROCEDE began with the *ejidos*. *Ejidos* participate on a voluntary basis, requiring a simple majority vote of approval from the *ejidatarios*. PROCEDE was launched in a campaign to promote a democratic and egalitarian relationship between peasants and the state, so that peasants themselves make the decisions that are important to their lives, reinforcing local governance. Rural promoters were employed to carry out the program at the field level, and attempt to convince the *ejidos* to enlist in the program. If the *ejido* agrees, the promoters then make preparations for measuring the land and resolving disputes (Appendini 2002). In Autlán, 44% of the land is owned by *ejidos* and 62% of *ejidos* had participated in the PROCEDE program by 2004 (Martinez 2004). The importance of land tenure with regard to land use will be further discussed in *Chapter 7*.

2.2 The municipality of Autlán

The municipality of Autlán covers 927 km^2 (92'732 ha) and has a total population of 50'834 (INEGI 2005). It is located at latitude: 19^{0}46'15" and longitude: 104^{0}22'10". The topography is very irregular due to its location in the *Sierra Volcánica Transversal*. In the northwest and southwest, mountains reach 1500-2700 m.a.s.l. An alluvial valley at an altitude of 900 m.a.s.l. dominates the center of the municipality. The remaining area can be described as hilly, with altitudes between 900-1500 m.a.s.l. Annual mean temperature is 22.5^0 C and annual mean rainfall reaches 732 mm/year in Autlán (INEGI 2005). The rainy season starts in May and ends in September. However, light rains can continue until the end of October.

Environmental conditions are very heterogeneous, yielding great diversity in vegetation. In the highlands (2300-2860 m.a.s.l.), fir-pine-oak forests can be found in humid sites, pine forests in rocky soils and broadleaf forest in small valleys. In the lowland (1000-1500 m.a.s.l.), mountain slopes are covered by deciduous oak forests. The small intermontane valleys with deep soils are dominated by agricultural fields and secondary vegetation. Parts of the plateau with steep slopes and shallow soils are covered with tropical deciduous forests and bamboo thickets. Another landscape type is the alluvial plain, which is dominated by intensive agriculture (e.g. sugar cane, tomatoes) and relicts of riparian forests and sub-deciduous tropical forests (Jardel et al. 1996, Vazquez et al. 1995). Traditional land use was cultivation of maize and beans, which changed to cotton and tobacco around 1950 with the arrival of American cotton and tobacco companies. This change was followed by the cultivation of vegetables and fruits, mainly melons for export to the United States. In 1975, a sugar refinery was constructed in the valley dominated by irrigated agriculture, and since then the cultivation of sugar cane is the predominant land use in the valley.

The study area, with the exception of a few families, is characterized by poverty, high illiteracy rates, minimal education, poor sewage systems and intra-familial violence. Maize cultivation continues to be important, but market prices have been low during the last decades. Other important crops include sugar cane, chili peppers, agave (for mezcal and tequila production), and tomatoes, all of which are grown mostly for export with the exception of sugar cane. Cattle breeding has gained in importance since the 1970s and is considered today to be one of the most important driving forces of land use change in the region. Farmers in the Ayuquila river basin in Jalisco have cultivated traditional green agave for at least a century. However, since 1996 the cultivation of blue agave (*agave azul*) in monoculture for use in the tequila industry has increased dramatically. Due to the fact that local farmers often lack the capital necessary to cultivate agave on their own, many farmers lease their land to independent contractors working for large tequila companies such as José Cuervo. The transfer of control of land from smallholders to contractors has potential social, ecological, and economic consequences mainly resulting from soil degradation caused by agave cultivation.

On the political-institutional level, the main problems include discontinuity of municipal governments, the multitude of government programs with distinct administrations (e.g. the 57 separate programs for sustainable rural development and agriculture), widespread corruption, and a lack of financial, human and technical resources to implement existing policies. The implementation of environmental policies in particular is weak, as priority is given to economic development. In addition, economic development and environmental conservation policies often contradict each other (Simonian 1995).

Figure 2-02 Map of the state of Jalisco

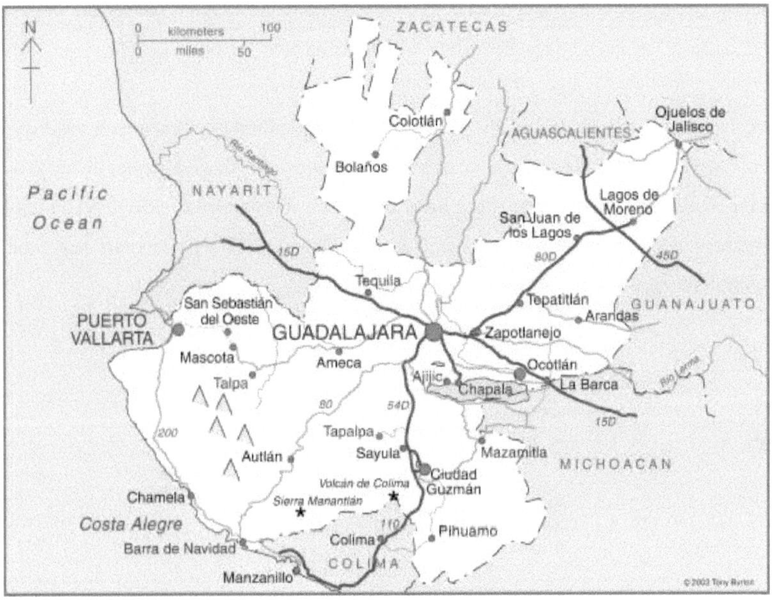

Source: www.mexconnect.com/mex_/chapalamap.htm

2.3 Case study sites

During the first phase of fieldwork, key persons from academia, private industry, municipal and state government and land managers in the study area were interviewed. A first analysis of the conducted interviews indicated potential driving forces and enabled the selection of the case study sites. Case study sites were selected according to the following criteria: 1) Land use type (agave or not, irrigation or not), 2) Land tenure status (agricultural communities and private landowners), 3) Geographical distribution in the municipality of Autlán to achieve a level of representativeness, and 4) Access to sites (accessible by public transport). Case study sites were selected to provide different contexts in which to examine land use/cover changes.

Three agricultural communities

The study was conducted in three agricultural communities and on three private properties in the municipality of Autlán. El Jalocote has 1'081 ha of land, Chiquihuitlán 13'663 ha and Mezquitán 1'159 ha. Autlán is located in the state of Jalisco in western Mexico (see **Figure 2-02**). El Jalocote and Chiquihuitlán can be reached by car within 40 and 20 minutes respectively from the medium-sized town of Autlán (39'310 inhabitants) (INEGI 2000). Mezquitán is situated 20 minutes from Autlán on the main road connecting Autlán to Guadalajara. Guadalajara is the second largest city of Mexico (5 million[5]) and can be reached from Autlán in 3 hours.

The three communities consist mainly of smallholder agriculturists who own a relatively small amount of land and typically engage in mixed subsistence and market strategies of production. Their main sources of income stem from agriculture, subsidies from the PROCAMPO program,[6] and remittances. A few families also engage in small businesses or have temporary part-time jobs in the horticulture or construction sector.

---

[5] http://www.usembassy-mexico.gov/guadalajara/Geliving.htm
[6] PROCAMPO is a government program that started in 1994. It can be translated as *Program for Direct Assistance in Agriculture*, and its main characteristics are the disbursements of payments to eligible farmers based on area planted, under the condition that farmers use their land for legal agriculture or livestock production, or for an environmental program.

Figure 2-03 Municipality of Autlán (State of Jalisco, western Mexico)

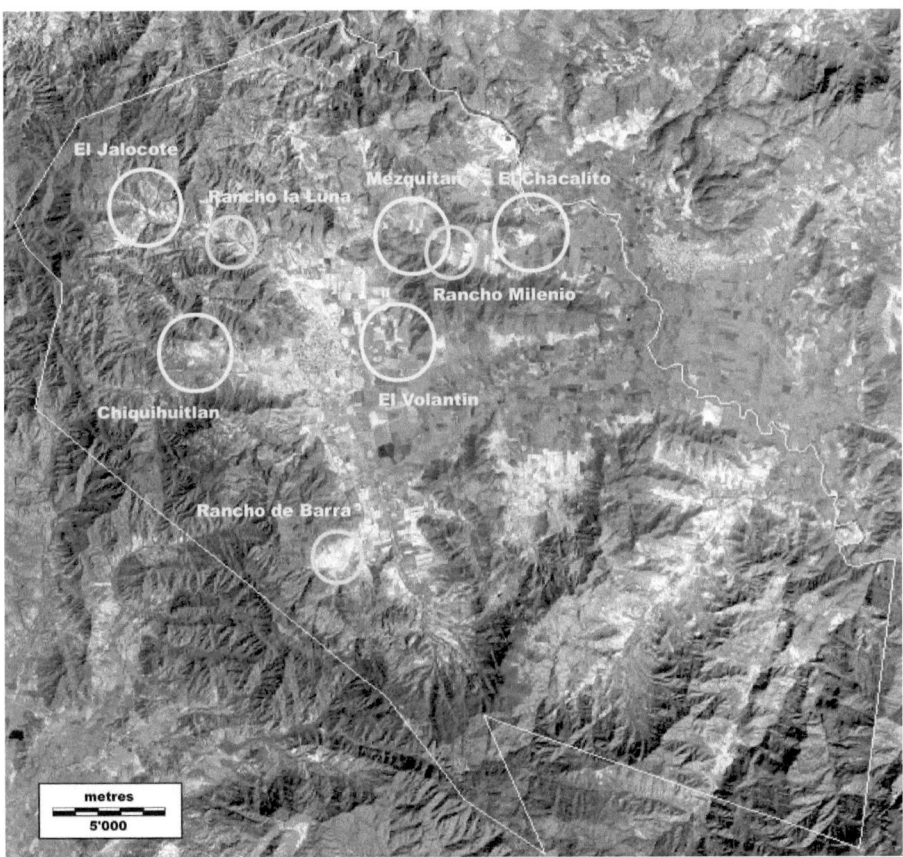

Composite RGB of Landsat bands 432 from 2000 with the municipality of Autlán outlined. In this picture (originally) red areas indicate higher levels of humidity (e.g. irrigated agricultural fields and forest). Circles represent case study sites; large circles indicate agricultural communities (El Jalocote, Chiquihuitlán, Mezquitán, El Volantín, Chacalito) and small circles indicate private properties (Rancho Milenio, Rancho la Luna, Rancho de Barra[7]).

---

[7] In order to ensure confidentiality, names of land owners and private properties, often called "Rancho ..." have been changed.

Chiquihuitlán is located west of Autlán at 1073 m.a.s.l. at the end of a valley. Regular bus service links the community to Autlán within 20 minutes. Chiquihuitlán has mainly dirt roads, a primary school, a kindergarten, a health clinic and a little shop. The village has no irrigation water, and drinking water comes from wells. The fields on the valley floor are almost exclusively cultivated with *agave azul*. The hills that rise around the village are covered with grass, patches of tropical deciduous forest and cacti, and are mainly used for pasture. An important source of income in Chiquithuitlán is the seasonal collection and selling of the prickly pear (*pitayas*) which is the fruit of the *nopal*, a typical local cactus. The fruits are considered a delicacy and fetch a high price. Most people in Chiquihuitlán either work on their own farms or work as day laborers in the regional agricultural and construction industries. Chiquihuitlán has 178 inhabitants (INEGI 2000). Chiquihuitlán is a *comunidad agraria*. The difference between an *ejido* and a *comunidad agraria* lies in its origin and land ownership status. Between 1917 and 1992, land was distributed to *ejidos* (a group of landless farmers requesting land) by the Mexican government. In contrast, a *comunidad agraria* is a group of people that have been living for hundreds of years in the same place, and are therefore considered the legal owners of the land and its natural resources.

Houses in Chiquihuitlán. Communal land in the hills can be seen in the background. Photo by author (April 2004).

El Jalocote is located northwest of Autlán at 1349 m.a.s.l. on hilly terrain and has 201 inhabitants (INEGI 2000). There is one cobblestone street, a health clinic, a kindergarten, a primary school and two little shops. It can be reached from the next town in 40 minutes by following a dirt road leading up the valley. The road was built in 1987, before which it was only a path and people traveled by foot or on horseback. Since 1992, a bus drives three times a day back and forth between the town of Autlán and El Jalocote. Most families are *ejidatarios* and own a field of irrigated agriculture close to the village and another piece of land in the hills that they use or rent out as pasture. Around 30 people from the community are working in the town of Autlán. Many women work as domestic employees or in restaurants, while the men work in the agriculture and horticulture industries picking fruit and cutting sugar cane. They travel each day to the Autlán-El Grullo valley, 40 minutes each way. Approximately 30% of the houses do not have running water or electricity. Many families from El Jalocote produce charcoal from the forests located in the hills. The charcoal is sold to traders that come to El Jalocote and then sell it in the region. El Jalocote has 5 permanent streams which provide some of the drinking water for the town of Autlán, a fact that is often mentioned with pride in El Jalocote.

A house of the *ejido* El Jalocote. Communal land in the hills can be seen in the background. Photo by author (March 2004).

Mezquitán is located at 939 m.a.s.l. on flat terrain on the main road linking the second largest city of Mexico (Guadalajara) to the Pacific Coast. Mezquitán has 773 inhabitants (INEGI 2000). It can be reached in 20 minutes on this road from the next town, Autlán. There is frequent bus service between Autlán and Mezquitán, around one bus every hour. There is a health clinic, kindergarten, primary school, secondary school, several small shops and one large restaurant. Almost all houses have running water and electricity. The *ejido* of Mezquitán includes three villages: Mezquitán, El Volantín and Chacalito, located in the north-central part of the municipality. While Mezquitán has no irrigated fields, all fields in Chacalito are irrigated and cultivated in sugar cane. In El Volantín, about 30% of the fields are irrigated and cultivated with maize or vegetables. In Mezquitán, a large number of rainfed fields are cultivated with *agave azul*. The *Commission Federal de Electricidad* (Federal Electricity Company) employs many people of the village of Mezquitán, contributing to the relative wealth of the *ejido*.

The main square (*plaza*) in the *ejido* of Mezquitán. Photo by author (May 2004).

Rincón de Luisa, case study site of Portner (2005): Located at 1196 m.a.s.l. Fields of the *ejido* are situated in the valley of Autlán and are irrigated and therefore almost exclusively used for sugar cane as it is the most valuable crop. Hills belonging to Rincón de Luisa are covered with forest and patches of grass, but only a few households use it for pasture in the rainy season. Some families collect cactus fruits in the dry season. Illegal plantations of marijuana are said to exist on top of the hills. Trickle irrigation was installed on the upper fields in 2000. Since then, the majority of households have sold their cattle and begun cultivating sugar cane.

La Laja de Abajo, case study site of Portner (2005): Located at 1000 m.a.s.l. Fields are all situated in the hills with no possibility for irrigation. Downhill, where slopes are not too steep, the company *Agave Azul y Servicios* rented all agricultural land to plant agave. Individual contracts have a duration of seven years, which corresponds to one growing cycle of *agave azul* from planting to harvest. Further uphill, farmers cultivate maize and beans and in the dry season use maize stubble and grass patches as fodder for their cattle. Some families practice shifting cultivation. In the rainy season cattle graze in the hills. Almost all families have members with jobs outside of agriculture.

Three private properties

Rancho Milenio: Land owner is 56 years old. Owns 100 ha of land, purchased in 1998. He plants 16 ha of agave, a few hectares of maize for cattle feed, and the remaining land is mainly used as pasture for his 60 head of cattle. He also owns a few horses and a large plot is dedicated to vegetables. He owns a business in telecom cables in Guadalajara and employs a land manager who oversees his ranch. He has never migrated to the United States.

Rancho la Luna: Land owner is 58 years old. Manages 70 ha of land since taking over the farm from his father. The land has been in the same family for several generations. Cultivates 18 ha of agave and the rest is used as pasture for approximately 50 head of cattle. The family also owns a butcher shop in the town of Autlán. No member of the family has migrated to the US.

Rancho de Barra: Land owner is 42 years old. Manages 44 ha of land since 1993 and owns 40 head of cattle. Cultivates maize and millet. Ranch is located close to the main road linking Guadalajara to Barra de Navidad on the Pacific coast. Owner has spent several long-term migration periods in the United States. Since 1994, he works every year for 6-8 months in the US and has opened his own business there (beauty salon).

3. Theoretical approach

Land use changes reflect how households are responding to their changing needs and goals and how they adapt to changing environmental, socio-economic and political conditions (Lambin et al. 2003). Changes in political, institutional and economic conditions can cause rapid changes in the rate or direction of land use change (Lambin et al. 1999). This study transcends the boundaries traditionally separating the natural and social sciences in order to provide a comprehensive analysis of the relevant phenomena. Land use change analysis is, by nature, interdisciplinary, and several different analytical frameworks have been developed to incorporate a range of theoretical and methodological approaches (Agarwal et al. 2002, Briassoulis 1999, Bürgi et al. 2004, Geist and Lambin 2002, Lambin et al. 2003). The interdisciplinary approach of studying land use changes has been driven by the request of the international research community for integration of results from various disciplines into a better understanding of what land use is, what drives land use changes and the potential socio-economic and environmental impacts of those changes.

3.1 The Land Use and Cover Change (LUCC) project

Much of the research on land use and cover changes has been undertaken in the framework of the Land Use and Land Cover Change (LUCC) project[8] of the International Geosphere-Biosphere Program (IGBP) and the International Human Dimensions Program on Global Environmental Change (IHDP) (Lambin et al. 1999, Turner et al. 1995). LUCC had a duration of ten years (1995-2005). Since 2005, research on land use and land cover change has been organized under the framework of the new 'Global Land Project'[9] which is located in the Department of Geography at the University of Copenhagen. LUCC defined itself as 'an interdisciplinary project/program designed to improve understanding and projections of the dynamics of land use and land cover changes as inputs to and consequences of global environmental change and as elements of sustainable development.' In order to do this, LUCC required new integrated global and regional models, informed by empirical assessments of the patterns of land cover change and by comparative case studies of land use processes, and was based on data and classification development. Furthermore, it also required major improvement in understanding how processes of land use and land cover change vary across spatial and temporal scales. The research agenda of LUCC was composed of five major research themes: 1) Integrated global and regional models, 2) Land cover patterns, 3) Land use processes, 4) Database and classification development and 5) Cross-scalar or scalar

---

[8] www.geo.ucl.ac.be/LUCC/lucc.html
[9] www.globallandproject.org

dynamics (Turner et al. 1995). The LUCC program was based on an explicit statement by the global change community that: 1) Global environmental change involves far more than potential climate change or loss in biological diversity worldwide; 2) Human agency and societal structures operate synergistically in complex ways with the environment to create this change and the responses to it and; 3) An improved understanding of the dynamics involved, with their implications for sustainability, requires a research strategy far wider ranging than that which has typified the history of the research community at large (Turner 1997). Since its beginning a large number of researchers have linked their work to the LUCC program, and this study was one of many projects endorsed by LUCC.

The Global Land Project builds on the research of the LUCC program. It has three main thematic areas: 1) The dynamics of land system change; 2) The consequences of land system change and; 3) Integrating analysis and modeling for land sustainability (Global Land Project 2005). One of the main recommendations made to the GLP is to investigate the role of remittances in relation to land use (Global Land Project 2005, Laumann 2006).

3.2 The impact of remittances

Much of the discussion on the development - migration nexus centers on the potential role of remittances in development. People migrate in order to ensure their own survival and to improve the economic wellbeing of their families. Once in their new host countries, migrants often send money home to provide for the basic needs of their family or to support community development projects. In the past 10 years, these migrant financial flows, known as remittances, have doubled worldwide and constitute the fastest-growing and most stable capital flow to developing countries (Kapur 2004). Remittances can be divided into private and collective remittances. Private remittances are funds sent directly from individual migrants to a family member in their country of origin. Collective remittances are sent by a group of migrants to their community of origin to support community and development projects or to contribute to disaster-related recovery efforts. Collective remittances represent an unspecified but relatively low share of total remittances (Goldring 2004). Remittances are now a key macroeconomic factor in the so-called Third World. For many developing countries, remittances are comparable to, or greater than, total export earnings, official development assistance, and foreign direct investment (Gammeltoft 2002) and have the potential to become the largest source of foreign exchange earnings (Glytsos 2002, Seddon 2004). In Mexico, remittances represent the second largest source of foreign exchange earnings after petrol (Faux 2006).

However, the development impact of these remittances is highly debated (Martin and Taylor 1996, Massey et al. 1998). As McDowell and de Haan (1997) show, those who argue for a positive impact suggest that by alleviating unemployment and providing strategic inputs such as remittances (Durand et al. 1996, León-Ledesma and Piracha 2004) and the skills of returning migrants (Olesen 2002), migration spurs development, narrowing regional disparities and eventually becoming unnecessary. While the initial increase in development may perpetuate migration to ensure continued growth, after a certain point, migration should decline as people have the opportunity to participate in local development (Nyberg-Sörensen et al. 2002). Critics, however, question whether migration, remittances and return are indeed directly converted into accelerated development (Seddon 2004). Contrarily, they might even create new dependencies undermining development both at national and regional levels, and thus perpetuating the North-South divide (Ellerman 2005). A third position sees migration more as a symptom of development rather than the result of a lack of it: "...*international migrants do not come from poor, isolated places that are disconnected from world markets, but from regions and nations that are undergoing rapid change and development as a result of their incorporation into global trade, information and production networks. In the short run, international migration does not stem from a lack of economic development, but from development itself.*" (Martin 1994, in: Massey et al. 1998).

A number of case studies at the micro-level have contested the view of uniform migration outcomes in developing countries, and have shown that these vary enormously even within communities (Basok 2003, de Haan et al. 2002, de Haan and Rogaly 2002, Gundel 2002, Mosse et al. 2002, Murray-Li 2002, Thieme and Wyss 2005). Nonetheless, it is generally recognized that migration has a largely positive economic impact on sending communities. Migration is no longer simply seen as a failure of development but increasingly as an integral part of the development process, with a potentially important role to play in the alleviation of poverty (Riak-Akuei 2005, Sander 2003). Adams and Page (2005) analyzed datasets on international migration, remittances and poverty for 71 countries and found that both international migration and remittances significantly reduced the level and depth of poverty. Households with migrants are in a substantially better situation than those who have no members abroad. Migration represents a form of insurance and results in families having diversified livelihood strategies and minimizing risk (Kapur 2004, Sorensen and Olwig 2002). In some countries such as Bangladesh, remittances account for more than half of the household income of migrant families, and in Senegal remittances comprise as much as 90% of household income (Sander 2003). Remittances have a positive impact on family wellbeing by enabling better health care, nutrition, housing and education (Seddon 2004).

Critical voices on the role of remittances point out that a culture of dependence might set in for receiving communities, with household members preferring to wait month after month for remittances instead of taking up jobs (Kapur 2004). Others are concerned about the potential of remittances to create inequalities (de Haan and Rogaly 2002) and finance "unproductive" spending, although the distinction between productive and unproductive spending is admittedly difficult to make (Seddon 2004, Verduzco and Unger 1998). A study on Mexican-US migration established that rural households use remittances for consumption, which is believed to have a strong multiplier effect on the national economy. Urban households however, were found to invest remittances in internationally produced items so that the money quickly leaves the country (Tienda et al. 1997). The largest concern is for the social costs of migration. While remittances constitute a crucial source of income for many families, the human costs are high in terms of uprooting, the breakup of families and the large burden of responsibility often placed on women that are left behind to care for the family (Banerjee et al. 2002, de Haan 1999, Kanaiaupuni 2000, Kannan 2005).

It has been found that remittance flows are more equally distributed (Jennings and Clarke 2005) and have lower transaction costs than foreign aid (Kapur 2004, Nicholson 2004). According to the World Bank, official international remittances sent home by migrants to approximately 80 countries represent the second most important source of external funding in developing countries (World Bank 2004). Estimates of remittances vary between USD 93 billion (Ratha 2004) and 167 billion (Wanner 2006) per year, which is at least double the level of official aid-related inflows to developing countries (Adams and Page 2005, Wanner 2006).

In 2004, the top three 'developing country' recipients of remittances were India, China and Mexico, followed by the Philippines, Morocco, Serbia, Pakistan, Brazil and Bangladesh (García-Zamora 2006). The largest flow of remittances (USD 45.8 billion in 2004) was to Latin America and the Caribbean (García Zamora 2006). Including unofficial remittances, this amount might be as high as USD 74 billion (Seddon 2004). Migrant remittances to Mexico in 1996 amounted to USD 4.3 billion, which is 14 times the total sum of net foreign aid received (The Economist 2002). In 2001, remittances to Mexico increased to USD 9.9 billion (Levitt and Nyberg-Sorensen 2004) and reached USD 16.6 billion in 2004 (García Zamora 2006) and 20 billion in 2005 (Faux 2006).

3.3 Investment patterns of remittances

In view of the fact that migration is a complex and multi-dimensional process, it is useful to distinguish between different types of remittances. Levitt (1998) used the term 'social remittances' to describe the social practices, ideas and values brought to migrant sending areas through the migration process. Knowledge, skills and technology brought back by returning migrants have been called technological remittances (Nichols 2002). Yet others refer to political remittances, describing changes in political demands and practices due to migration. Durand (1994) differentiated remittances according to their function. The first type refers to wages that are sent home by circular migrants to support relatives. These migrants are typically from regions characterized by limited investment opportunities or low-yield agriculture. The second type refers to remittances as investment that are either sent home or brought back upon return. This type of remittance is typical of migrants that make a limited number of trips with a specific objective, such as saving money to buy land or construct a house. The third type refers to remittances as capital: funds saved specifically to invest in a business. This type of enterprise is the most difficult to achieve.

The socio-ecological context and circumstances of an individual household greatly influence how remittances are invested. But despite local differences, there are general patterns of investment that have been observed in different regions of the world. The prevailing notion is that 60-80% of remittance income is used to cover basic needs such as food, medical expenses and education, while the remaining 20-40% is invested in land, livestock, housing, business ventures and savings (Delgado and Rodríguez 2001, Goldring 2004, Massey and Parrado 1998, Nyberg-Sörensen et al. 2002, Wanner 2006).

Seddon (2004) reports this general investment pattern in South Asia: After debts are paid off (particularly those incurred in order to migrate), basic necessities including health costs are covered. Third, housing is improved and basic consumer durables such as radios are acquired. Fourth, an increasing priority is given to the education of children, to save funds for the migration of another household member and to maintain social networks. Only after all of these demands are met will the household invest in productive assets or some enterprise in agriculture, manufacturing or other sectors. Nyberg-Sörensen et al. (2002) indicate a similar pattern for investment of remittances in the developing countries they studied:[10] First priority is given to daily living costs and repayment of debts. Over time, remittances are invested in improved housing and education followed by 'conspicuous consumption' such as the purchase of radios and televisions. Finally, remittances are

---

[10] Studies were conducted in Least Developed Countries (LDCs), but the exact countries were not specified.

invested in productive activities, including improvement of land productivity. McKay (2005) reports a slightly different pattern from the Philippines, where remittances are spent on the construction and renovation of houses and corner shops, and the purchase of agricultural inputs and cars and motorcycles. Furthermore, returning migrants bring clothing and gifts and frequently send home large boxes of household goods. In some cases, remittances are used to start a small business such as a tailor or woodcraft shop. Remittances are also invested in cultural capital by paying for weddings, funerals and education. Sending family members to urban schools is also a preferred investment of remittances.

The overall sense is that migration leads to higher household income and reduces poverty levels, but does not generally lead to the type of economic development that would generate new employment opportunities and foster long-term economic growth (Bichsel et al. 2005, Durand and Massey 1992). Remittances are considered to have a local-level growth impact if they generate jobs and diversify the economy. Durand et al. (1996) suggest that remittances stimulate economic activity both directly and indirectly. They argue that the spending of remittances leads to significantly higher levels of employment, investment, and income within specific communities as well as at the national level. Even though two-thirds of remittances are spent on consumption, this spending has a strong multiplier effect on the Mexican economy. For instance, remittances are spent on clothes, electronics, doctors, and pharmacists. As these funds work their way through the economy, they steadily multiply to increase income, production and investment. Similar results from a study in Kerala, India showed that the spending of remittances on consumption led to economic growth. High levels of spending on telephone conversations with migrant family members had a strong multiplier effect on the local economy as the telecommunication sector expanded and infrastructure increased (Pushpangadan 2003).

While spending patterns of remittances are similar for many developing countries, the specific implications of remittances for land use, land cover and landscapes are disputed. Some claim that migration leads to greater investment and agricultural improvement, while others claim that it leads to a loss of labor and degradation of agricultural systems. The lost labor effect is assumed to have a negative impact on local production by causing a critical shortage of agricultural labor. This loss of labor can cause a de-intensification of agriculture and the decline of land area under cultivation (de Haas 2005). Furthermore, labor scarcity may have a deleterious effect on the cultural and social organizations that sustain agriculture (Jokisch 2002).

However, the effect of labor loss on agricultural production may be countered by the investment of remittance income in agricultural activities. Even though a relatively small amount of remittances is spent on productive activities, there are a number of studies from Latin America, West Africa and South Asia which show that remittances are used to improve agricultural land (de Haas 2005). Remittances allow migrant households to purchase improved inputs, increase yields, grow market crops, expand irrigation, and hire labor, as well as to overcome capital and credit constraints. In the town of Alvaro Obrégon in Mexico, it was found that 30% of remittances were spent on land, tools or livestock (Trigueros and Rodriguez 1988).

In certain areas in India, significant portions of remittances are invested in production-increasing technology in the agricultural sector (McDowell and de Haan 1997). In Kenya, the recovery of Akamba lands in Machakos from soil erosion was largely achieved through out-migration of Akamba men and the investment of remittances into soil conservation (McDowell and de Haan 1997, Tiffen et al. 1994). Some experiences from West Africa and the Maghreb region indicate that a considerable proportion of remittances are used for investment in agricultural land, equipment, and small-scale businesses. For instance, remittances have been used to finance irrigation and other agricultural inputs (Nii Addy et al. 2004). An interesting study from the Philippines described how female migration transformed the landscape in their community of origin from subsistence rice production to commercial bean cultivation (McKay 2005). As the women are abroad and therefore not available for fieldwork, the men require new technologies and wage workers to replace them. The remittances are used as capital to purchase land and to cultivate new commercial agricultural crops in an attempt to diversify livelihoods and increase household security. Yet, migration is not only a cause but also a result of the ongoing agricultural transition, as profits from bean gardening allow additional family members to migrate.

While in some areas remittances are invested in equipment, seeds, fertilizer, and draught animals or hired labor (McDowell and de Haan 1997), in other areas this is not the case. In one study in Bangladesh, migrant families bought up agricultural land but left agriculture (Islam 1991). This meant that land use changed, land prices went up and agricultural production decreased. In the Sahel, several case studies showed that migrant families were not able to spend remittances on agricultural production because the costs to hire labor and buy agricultural materials and chemical fertilizer were so high that only very few migrant families were able to make the investment (David 1995). Similarly, remittances in Thailand are not invested in agriculture due to high costs for agricultural inputs and labor and low output prices, which often drives small farmers into debt (Jones and Pardthaisong 1999). In fact, as a consequence of receiving remittances, land was farmed less intensively and in some cases not worked at all. Land was often rented out to sugar cane growers, causing a land use change from rice to sugar cane cultivation. This trend accentuated migration movement because as less rice was grown for subsistence, dependency on remittances increased. These contradictory examples demonstrate that the relationship between agriculture and migration is complex and highly dependent on local circumstances, and raises the question: Under what circumstances are remittances invested in agricultural activities, or more generally, under what circumstances do remittances influence land use?

3.4 Factors determining investment patterns of remittances

As outlined in the previous section, investment of remittances is influenced by a number of factors, which will now be discussed in more detail. One factor is the type of migration. Temporary,[11] circular,[12] and permanent[13] migration each have different impacts (Verduzco and Unger 1998). The following study illustrates well the differential impact remittances can have within the same community based on the type of migration. Mines and de Janvry (1982) found that long-term Mexican migrants aiming to stay in the US bought up village land with remittances, thus changing local production systems from staple crops to low productivity cattle-raising. Since a semi-skilled job in the US provides much more income than does improved land in the village, the land was not intended for their own use. Instead, land was purchased to provide their parents with a source of income and as security in case they had to return from the US. Temporary migrants, on the other hand, would have liked to invest in improved land but lacked sufficient resources. Instead, temporary migrants maintained land in traditional corn production in order to ensure their children a source of income in case they failed to find work in the US (Mines and de Janvry 1982).

As patterns of remittance spending at the household level influence long-term beneficial impacts, gender differences in spending need to be taken into account. Women often invest more in healthcare, food and education while men tend to purchase conspicuous consumer goods such as televisions and cars (Sander 2003). These differences in spending patterns have potentially long-term effects for migrant families, as education and health care have investment effects which might help raise households out of poverty where conspicuous consumption would not (Kannan 2005,

---

[11] Temporary migration refers to persons who are very likely to return to country of origin after a limited stay, without acquiring legal permanent resident (immigrant) status.
(www.census.gov/population/www/documentation/twps0060.html).
[12] Circular migration is short-term (1-6 months), often seasonal, labor migration that does not involve a permanent change in residence (Chapman and Prothero 1985).
[13] Permanent migration refers to persons having obtained residence, citizenship or employment on a permanent basis in the destination country (Alfieri and Havinga 2005).

McKay 2005). Another factor in spending patterns is related to the size of remittance, as migrants who remit larger amounts have been observed to invest more in productive activities (Jones 1995, Massey and Basem 1992). Durand and Massey (1992) found that the age of migrant stream, the quality of local resources, the industrial structure in the US destination region and the specific life-cycle stage of a household were important factors influencing the investment of remittances. With regard to life-cycle stage, they observed that during the family formation stage, most earnings are spent on housing, health costs and recurrent household expenses. As the migrant household ages, a higher percentage of earnings are invested in productive activities. It has also been suggested that migrants are more likely to invest remittances into productive activities if they already have a certain threshold level of economic resources such as land or a small business (Massey and Basem 1992). These can then be expanded and reinforced with the help of remittances. Finally, patterns of remittance spending also depend on the availability of opportunities for investment. Sending communities may be in a geographic location that is unsuitable for productive investment. Migrants often come from small rural villages located far from markets and lacking in basic infrastructure such as paved roads, electricity, running water and sewage. In Mexico, 80% of municipalities with high migrant activity have less than 20'000 inhabitants and only one municipality has more than 50'000 inhabitants. This indicates that migrant sending areas are predominantly rural (Verduzco and Unger 1998). Other communities are not suitable for investment due to poor quality land, lack of water, or land tenure systems which preclude the purchase of land (Durand and Massey 1992). In Mexico, the purchase of land was possible until the 1930s but became increasingly difficult as private land was redistributed to *ejidos*. Even though sales of *ejidal* land took place throughout the 20th century, this only became legal again in 1992 (Appendini 2002). Thus, a primary opportunity for agricultural investment was limited for decades by land tenure regulations.

De Haan (1999), while admitting the complexity of the relationship between livelihoods, migration and remittances, argues that if the right incentives for agriculture exist, migrants invest in agriculture in a rational manner. He concludes that the impact of migration on agriculture depends on a number of factors, namely: the context, type of migration, educational level of migrants, length of stay, assets, and local social structures and institutions.

Jokisch (2002) argues that it is the political, economic and environmental context from which migrants leave that determines how remittances transform the landscape. Thus, favorable economic, political and biophysical conditions encourage investment in agricultural production. This was also found by Durand and Massey (1992) in their review of the impact of Mexican migration: productive (agricultural) investment tended to occur where political and economic incentives and environmental conditions were favorable, but usually after subsistence needs were met. De Haas (2005) follows the same line of argument: *"Migration and remittances can potentially contribute to development, but the specific political, economic and social circumstances in both the sending and receiving countries determine the extent to which this potential is exploited."* These propositions are along the same lines as new economic labor migration (NELM) theory (Taylor 1999). NELM theory suggests that the demand for migrant remittances increases as development proceeds and both investment opportunities and returns on investment improve. In contrast to negative views of the impact of migration on development, NELM theorists argue that migration sets in motion a development process, diminishing the production and investment constraints faced by households in imperfect market environments and creating income growth. Remittances thus play a positive role in economic development (Taylor 1999). NELM theorists argue that rural communities located in areas with high-quality land, better infrastructure, and good access to markets offer enhanced opportunities for profitable investment and therefore are more likely to invest in productive activities (Basok 2003). On the other hand, migrants coming from areas with poor-quality land and limited infrastructure would be expected to spend remittances primarily on daily household needs. However, a number of studies have contradicted the predictions of NELM theory. Jones (1995) found that migrants in a dynamic region of Coahuila in Mexico invested mainly in consumer goods whereas migrants from the less developed region of Zacatecas invested heavily in agricultural inputs. Basok (2003), in a study on investment patterns in 11 Mexican villages, found that while migrants from better endowed communities did invest more in businesses than those from less well-endowed communities, the opposite was the case concerning investment in land purchase. Basok argues that the price of agricultural land is the decisive factor behind land purchase. In communities with good market access and infrastructure, agricultural land was most expensive and therefore much harder to purchase. In the worst endowed communities, land was much cheaper and therefore more easily acquired.

3.5 Migration concepts and definitions

Migration research attempts to explain the causes of international migration as well as social, cultural, economic or political consequences for the destination and sending region (Deshingkar and Start 2003). These analyses are often conducted in the traditional framework of push-pull factors, where push (e.g. poverty, lack of work) and pull (e.g. better salaries in destination countries) factors that influence migration flows are present in the sending and receiving areas (Thieme 2006).

Potential impacts of migration depend on the type of migration (McDowell and de Haan 1997). Migration can be classified according to spatial, temporal or causal criteria. According to spatial criteria, a distinction is made between internal and external (international) migration. Furthermore, rural-rural is distinguished from rural-urban migration. Under temporal criteria, the types of migration include temporary, circular and permanent migration. Finally, causal criteria refer to the reason for migration, which is often categorized as voluntary (e.g. in search of employment) or involuntary (e.g. refugees of war zones) (McDowell and de Haan 1997). In addition, research on migration patterns has shown that the social experience and consequences of migration are not uniform, but are influenced by class and gender. Patterns of movement are shaped by context-specific dynamics, and mediated by social networks, gender relations and household structures (De Haan et al. 2002). Kaufmann et al. (2004) argue that spatial mobility, such as migration, can be considered as a type of capital that is related to, and influences, social mobility. Therefore, patterns of migration vary between social groups in a particular location and also between households within the same social group.

In the framework of this study, the focus is on the impact of international labor migration on land use in the sending areas. An international labor migrant is "a person who is to be engaged, is engaged or has been engaged in a remunerated activity in a State of which he or she is not a national."[14] In the context of Mexico-US migration, international migration can be described as 'transnational migration.' The term 'transnational migration' emerged in the early 1990s when scholars encountered migration practices outside the bounds of conventional migration theories. To overcome the dichotomy of migrants that either depart (emigrants) or arrive (immigrants), the term transnational migration emerged to describe a situation in which sending and receiving societies are understood to constitute a single field of analysis (Levitt 2001, Levitt and Nyberg-Sorenson 2004). Transnationalism describes a process by which migrants build and sustain social relations that bind together their societies of origin and settlement (Bailey 2001, Basch et al. 1994, Vertovec 1999). Transnationalism can be defined as "occupations and activities that require regular and sustained social contacts over time across national borders for their implementation (...) and it involves individuals, their networks of social relations, their communities, and broader institutionalized structures such as local and national governments" (Portes et al. 1999). Taking a transnational approach shifts the analytical focus from a fixed place to a mobile process, and from 'place of origin' and 'place of destination' to the movements involved in sustaining cross-border livelihoods (Sorensen and Olvig 2002). Contrary to conventional migration theory, transnational approaches

---

[14] UN-Convention on the Protection of the Rights of All Migrant Workers and their Families, Article 2, 1990 (cited in: Thieme 2006).

suggest that migration should be understood as a social process linking the sending and receiving countries (Nyberg-Sorensen et al. 2002). According to Glick Schiller et al. (1992), migrants are designated "transmigrants" when they develop and maintain multiple relations – social, economic, political, organizational and religious – that span borders. According to Portes et al. (1999) and Landolt (2001), transnational migration is established when:

- A large number of migrants engage in transnational practices (a mass phenomenon)
- Transnational practices are of a certain stability and resilience over time, and
- These transnational activities are not captured by an existing concept.

These conditions appear to be met for Mexican migration to the USA. However, according to the following definition of transnationals, not all Mexican migrants from the study site could be considered transnationals. "Labor migrants become transnationals when they or their families have multiple home bases and ongoing commitments and loyalties that straddle political territories" (Aymer 1997). Nonetheless, the term 'transnational migration' will be used here as it succinctly describes the migration processes for the majority of migrant families in the study site.

3.6 Theory of land use change

Linkages between population and nature have been the focus of a number of theories. Two important historical theories for framing the debate over the impact of population growth are those developed by Malthus (Meadows et al. 1971) and Boserup (1965). Malthus pointed out that while food production levels increase at a linear rate, human population increases at a geometric rate if unchecked. Therefore, Malthus' theorem states that as population continues to increase, the decrease of available food per capita will lead to famines and result in the extinction of the human race. Boserup proposed that it is not agricultural methods that determine population (via food supply) but population that determines agricultural methods. She showed that increased population leads to agricultural intensification. Moving beyond the focus on population growth, Blaikie and Brookfield (1987) have shown that environmental change can occur under expanding, declining or constant population levels. Environmental change has also been examined in terms of single factors such as the economy (Mather et al. 1999) or multiple sets of factors (economy, institutions) (Lambin et al. 2003). Since land use analysis is not yet established as a science in itself,[15] studies of

---

[15] In 2006, the GLP launched a new journal named *Land Use Science*. It is based on the knowledge created by the land use research community (Turner 2001) and in particular by the LUCC program (1995 – 2005), and promotes the use of the term 'land use science': "The study of the nature of land use and land cover, their changes over space and time, and the processes that produce these patterns and changes can be termed

land use change are often based on general theoretical frameworks from other disciplines such as geography, economics and environmental science. As interaction between humans and the environment becomes increasingly complex, the analytical approaches are continuously revised (Veldkamp and Lambin 2001, Verburg and Veldkamp 2005).

It is helpful to go back to the original meaning of *"theory,"* From the Greek for *"looking at something"* or *"observing something"* (Briassoulis 1999). Therefore, theory is the basis for knowledge. It is *"a set of connected statements used in the process of explanation"* (Briassoulis 1999) or *"a system of thought which, through logical constructs, supplies an explanation of a process, behavior, or other phenomenon of interest as it exists in reality"* (Chapin and Kaiser 1979, p. 27 cited in Briassoulis 1999). Following this definition, a theory of land use change describes the structure of the changes from one land use type to another and explains why these changes occur. This includes explanation of what causes the changes and what mechanisms are at work.

So far, no general theory of land use change exists and it is questionable whether such a theory is desirable and possible. Whether or not it is desirable depends to a large extent on the epistemological perspective. Researchers adhering to idealism, postmodernism and realism who put great emphasis on the importance of context may not consider a general theory appropriate or useful. In addition, the diversity of contexts in which land use changes occur makes it unlikely that a general theory can be developed despite the existence of broad patterns and regularities over space and time. Furthermore, a general theory of land use change would mean the loss of the specific details of particular contexts, which often have critical explanatory power. The possibility of developing a relatively simple explanation of *"why we transform the environment the way we do"* remains a challenge (Scoones 1999, Turner et al. 1993).

Nonetheless, the role of theories of land use change is very important because theory guides policy. Inadequate theories of land use change have the potential to misguide policy and thereby cause damage instead of contributing to more sustainable natural resource management. One key challenge is to develop projections of how land use is likely to change in the future based on knowledge gained from the past and the present. The difficulty resides in the fact that systems are dynamic and underlying processes of change are likely to change themselves.

'land use science'" (Taylor and Francis 2006). It can be expected that publications in the journal *Land Use Science* will significantly contribute to theory-building.

This research adopts a realist epistemological perspective. Realist explanations use abstraction to identify the necessary/causal powers of specific structures which are realized under contingent/specific conditions. Realism regards the world as differentiated, stratified, and made up not only of events (as positivism does) but also of mechanisms and structures.. Structures are seen as sets of inter-related objects that have essential properties and hence characteristic ways of acting. That is, they possess "causal powers and liabilities" (Sayer 1984) by virtue of what they are and which are, thus, necessary. Realist analysis tries to identify causal chains which explain particular events in terms of the behavior of these structures (Briassoulis 1999).

This study takes a realist perspective on the understanding that theories can illuminate the past and the present and provide insights regarding future trends. However, realism does not believe in a universal theory that is able to predict the future. Acknowledging the diversity of real world situations, realist epistemology does not attempt to provide a universal theory of land use change. Land use change theory may provide a broad explanatory framework, including driving factors and patterns of change, but it does not provide those details, which may be critical in explaining land use change in particular contexts and circumstances.

3.7 Political ecology

The political ecology approach adopted for this research is used to investigate human-environment interactions in all their biophysical and socio-economic complexity (Greenberg and Park 1994, Soliva 2000). Interactions between the environment and the socio-economic sphere consist of dialectical, historically derived and iterative relations between resource use and the socio-economic and political contexts, which shape them (Blaikie 1999). *"Political ecology combines the concerns of ecology and a broadly defined political economy. Together this encompasses the constantly shifting dialectic between society and land-based resources, and also within classes and groups within society itself"* (Blaikie and Brookfield 1987). Political ecology develops the common ground where various disciplines intersect and is therefore based on a plurality of disciplinary backgrounds. The following paragraphs briefly outline the intellectual origins of political ecology.

In the 1950s, cultural ecology focused mostly on cultural adaptations to the environment (Bryant and Baily 1997). The notion of "ethnoscientific knowledge" is central to cultural ecology, meaning knowledge about the resource use strategies of indigenous subsistence communities who have non-Western agro-scientific knowledge (Peet and Watts 1996). Cultural ecology was criticized for being too simplistic; it was accused of portraying societies as a product of environmental circumstances, and of not paying sufficient attention to sociological factors (Schubert 2005). As a result, the 1970s saw the emergence of political ecology, with less of a deterministic bias. Political ecology focuses on the political issues of structural relations of power over environmental resources (Blaikie and Brookfield 1987). Political ecologists refuted most of the neo-Malthusian assumptions that the increase of human population will exceed food production levels, ultimately leading to famine and the extinction of the human race. Blaikie and Brookfield (1987) argue that "critical population density" is unlikely to exist for a certain piece of land, if carrying capacity of the land changes with new technology or in years of rich harvests (Schubert 2005). Blaikie (1985) puts forward a "chain of explanation" model, which examines how multiple processes are shaping resource use. Within this conceptual framework, land degradation is analyzed on the basis of causal chains between the "land managers" and their land, other land users, stakeholders from the wider society who affect them, the state and finally, the global economy. From the end of the 1980s to the mid-1990s, political ecology gained popularity more because of the analytical lens it provided than as an all-inclusive theory on human-environment interactions (Nüsser 2003, Schubert 2005).

One of the key debates in political ecology centers on whether the starting point of analysis should be humans and human agency or nature and biophysical dynamics. Scholars coming from the ecological sciences often feel that political ecology focuses too much on the social and political dimensions of resource access while neglecting the biophysical realities of the natural environment (Walker 2005). They argue that ecosystems have to be considered as active agents. Some, like Vayda and Walters (1999), even argue that most of what is done under the umbrella of political ecology is politics without ecology and should be termed 'political anthropology' or simply 'political science.' They suggest that one should begin by observing environmental changes and then move from there to seeking causes, rather than assuming that the most important causal factors are political and thereby potentially overlooking equally or even more important factors. However, others argue that the advantage of political ecology resides precisely in the fact that instead of having a decidedly natural science or social science approach, it focuses on the mutual constitution of social and environmental change (Derman and Ferguson 2000). Despite the fact that a number of scholars are contributing to the theoretical development of political ecology (Bryant and Bailey 1996, Forsyth 2003, Peet and Watts 1996) some scholars argue that the main weakness of political ecology consists of the lack of coherent theory-building.

For this research, political ecology was the most suitable approach for five reasons: 1) In political ecology the natural environment is perceived as a setting for human action, which - at the same time - is modified by such action, thus acknowledging the complexity of human-environment interaction;

2) It is well-suited for actor-oriented research; 3) Political ecology integrates historical context and investigates interactions at different scales, both of which are key factors in land use change; 4) It takes into consideration power relations between actors and; 5) It allows one to establish chains of explanation leading from the observed land use changes to the proximate causes and finally to the underlying driving forces. This research begins with the environment, investigating land use changes and then working backwards in time to construct chains of causes and effects. This approach, based on the "chain of explanation" model elaborated by Blaikie (1989, 1995), has several advantages. It does not put politics in the center of the investigation by searching for on-the-ground impacts of policies. This would be difficult, and assumes beforehand that politics are the most important factor to consider. Instead, it starts from observed land use changes and links these changes to the livelihood strategies of the land managers, in this case rural households. The chain of explanations then leads to those factors influencing practices of the land manager, such as land tenure systems, subsidies, and labor allocation. These factors operate at multiple scales, from the household to the community to the regional and finally the national level. The state in turn is part of the global economy and as such is impacted by world trade, foreign debt crises, trade agreements and structural adjustment policies. Here the causal chain leads from the local to the regional, the nation state to the international level. The interactions between these different levels have been conceptualized by Hurni (1998) in a multi-level stakeholder approach to sustainable land management. While this approach avoids the danger of overlooking important factors by focusing only on political aspects, the challenge lies in then being able to identify causal connections between observed land use changes and their drivers.

3.8 Research questions and hypothesis

Research aim: The research aim is to describe and analyze temporal and spatial land cover and land use changes at the municipal level in western Mexico. It aims to identify the driving forces for land use changes and develop an understanding of the processes that have caused these changes. In particular, the influence of remittances resulting from Mexican transnational migration to the US on land use changes is examined. The environmental effects of these land cover and land use changes are also evaluated.

The temporal and spatial delimitation of this thesis are land cover and land use changes that have taken place over a ten-year period (1990-2000) in the municipality of Autlán, Jalisco State, western Mexico. Availability of comparable satellite images with the same resolution of 28.5m x 28.5m determined the time period. The spatial delimitation - the municipality of Autlán - was determined by the location of a NCCR North-South research project in this same area.

Research questions

The present project addresses four main research questions:
1) What land use changes have taken place between 1990 and 2000?
2) What are the driving forces and dynamics underlying those changes?
3) What is the impact of migration and remittances on land use and cover change?
4) What are the differences in land use strategies between migrant and non-migrant households?

Even though the third research question will have a strong focus on the influence of the investment of remittances on land use change, it will also include an analysis of migration patterns, push and pull factors leading to migration and the effect of the loss of labor. In order to analyze the influence of remittances on land use and cover changes, the concept of "remittance landscape" will be used.

Hypothesis: Migration is a driver of land use change

In recent years labor migration has been increasingly recognized as an important livelihood strategy for many families in developing countries in coping with crop failure and poor market prices for agricultural products (Adams and Page 2005, de Haan 1999, de Haas 2005, Deshingkar and Start 2003, McDowell and de Haan 1997, Nyberg-Sörensen et al. 2002). The specific research area in Mexico is characterized by high rates of rural migration to the US. It is hypothesized that remittances influence land use change because remittances change the economic situation of the rural migrant households and that the investment of remittances influences land use change. Furthermore, it is hypothesized that migrant households use different land use strategies than non-migrant households.

4. Conceptual framework

Doing research in the field of land use/cover change, there is a certain danger of falling into the 'ideographic trap' of producing research findings valid only for a specific and spatially limited area. A trade-off has to be made between rendering research results comparable and avoiding unjustified generalizations. By choosing the extensively used conceptual framework of proximate and driving forces underlying land use and cover changes, research results presented in this thesis can be compared to other land use and cover change studies and contribute to theory-building on land use change.

4.1 The concept of proximate causes and driving forces

In 1998, a LUCC workshop concluded: *"the largest gap in land use and land cover change characterization is not the extent, pace and direction of land cover changes, which can be approached by remote sensing, but the functional understanding and adequate parameterization of land use dynamics, e.g. selections of variables able to characterize interrelations and interdependencies of the elements of a land use system, like land use purpose, land use interventions and their human and biophysical driving forces"* (Baulies and Szeijwach 1998). Research related to land use and land cover change therefore tries to understand what causes changes. Two main concepts, proximate causes and underlying driving forces are used to describe the factors behind change. These concepts have been developed primarily by Geist and Lambin (2002) and are now widely used in the LUCC research community (Agarwal et al. 2002). Proximate causes are human activities (e.g. cattle ranching) that directly affect the environment and thus constitute proximate sources of change, such as deforestation. Underlying driving forces are fundamental forces behind proximate causes of land use and land cover change. The underlying driving forces are social and biophysical processes that directly or indirectly affect the decision-making of the land user. A number of studies, such as Geist and Lambin (2002), have shown that proximate causes and driving forces of land cover could not be reduced to single variables. Nonetheless, the hypothesis has been put forward that the main driving force of land use and cover change are economic opportunities, mediated by institutional factors (Lambin et al. 2001, Meyer and Turner 1994). An improved understanding of land use and cover change dynamics is essential for responding to environmental change and for managing human impact on natural systems. This includes better knowledge on feedback mechanisms between land use and underlying drivers. For instance, what kind of feedback can accentuate the speed or intensity of land use change, or constitute human mitigating forces, for example via institutional actions that counteract factors or their impact? What are the central factors that influence positive and negative feedbacks in land use change? How do different political and institutional contexts affect feedback mechanisms in different land use and cover change scenarios? (Lambin et al. 2003).

Since the early 1990s, several different categories of driving forces have been used. For example, Turner and Meyer (1991) and Stern et al. (1992) use the categories of population, level of affluence, technology, political economy, political structure, and attitudes and values. However, with the growing body of research on land use and cover change a consensus has been established that the following six categories are most useful: economic, political and social institutions; culturally determined attitudes, beliefs and behavior; technology; demography; and environmental factors (see Agarwal et al. 2002, Bürgi et al. 2004, Geist and Lambin 2002). The challenge is to investigate the relative importance of these driving forces and the interactions between them. This analysis should allow for an assessment of future scenarios and the relative impacts of different policy choices. Furthermore, the use of generic categories for driving forces allows for a comparison between, and integration of, studies on land use change conducted at different temporal and spatial scales.

In the present study, land use and land cover at t0 (1990) was compared to land use and land cover at t1 (2000) using remote sensing and GIS technology. Next, proximate causes such as land tenure, agricultural subsidies, and the presence or absence of irrigation water were identified based on surveys and in-depth interviews. Third, proximate causes were then related to a number of underlying drivers which were classified according to the categories developed by Geist and Lambin (2002), listed above. The main hypothesis of this research is that migration influences land use. However, as migration is the result of socio-economic, political and environmental processes, conceptually it can be part of at least four types of driving forces: demographic, economic, political and social institutions and culturally determined attitudes and beliefs. This model is based on the assumption that land use at t0 feeds back to driving forces and proximate causes, which in turn influences land use at t1.

Figure 4-01 Visualization of conceptual framework

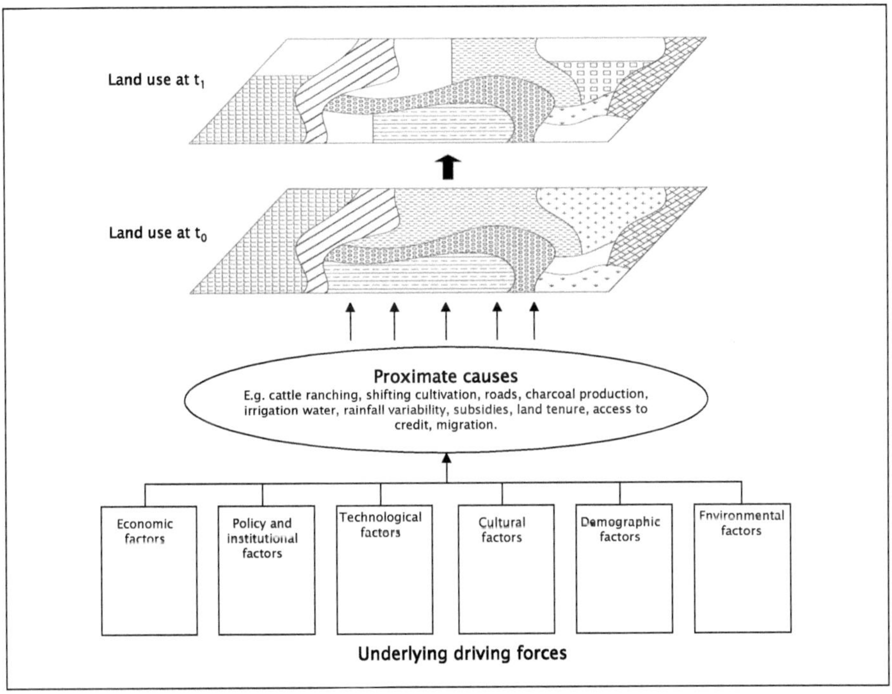

[Figure: Diagram showing two land use layers (Land use at t_1 and Land use at t_0) connected by an arrow, with Proximate causes (E.g. cattle ranching, shifting cultivation, roads, charcoal production, irrigation water, rainfall variability, subsidies, land tenure, access to credit, migration.) below, and Underlying driving forces: Economic factors, Policy and institutional factors, Technological factors, Cultural factors, Demographic factors, Environmental factors.]

Diagram by author.

4.2 The concept of remittances landscapes

Despite the impressive increase in remittances worldwide, and the substantial amount of research on migration, there is relatively little knowledge about the impact of remittances on landscapes. At present, no commonly used definition of remittance landscapes seems to exist. One possible definition has been developed by the PhD researcher in the framework of her participation in compiling an encyclopedia of land use and land cover change (Geist 2006). A remittance landscape is defined there as follows: *"An emergent landscape that is driven by the investment of remittances"* (Hostettler 2006). So far, the transformation of at least two types of landscape due to the investment of remittances have been identified.

The first type of transformation is that of an agricultural landscape to a peri-urban landscape. Investment of remittances in house construction replaces traditional dwellings with large brick and cement homes. Many migrants decide to leave their country in the hope of earning enough money in another country to be able to build a house in their community of origin (Durand and Massey

1992, Tiemoko 2004). This can lead to the emergence of peri-urban landscapes of *"cultivated real estate,"* when land is valued as a safe investment and a place to demonstrate one's financial success by building a large home, rather than as an agricultural investment. This type of remittance landscape dominated by large ostentatious houses to demonstrate achievement and to provide a place for retirement has also been described as a *"landscape of conspicuous retirement"* (Jokisch 2002). In a few migrant sending areas, this transformation includes the partial abandonment of cultivated land, because remittances substitute for agricultural income and are sufficiently large to allow households to abandon an economic livelihood based on agriculture, for example in Bolivia, Ecuador, Mexico, Philippines, Thailand, and India (Jokisch 2002, Lambin et al. 2001, Nii Addy et al. 2004). In the African context, even though remittances are also invested in housing, meeting basic needs continues to be the central preoccupation (Gundel 2002, Sander and Munzele 2003, Tiemoko 2003). Nevertheless, some migrants are able to build homes for their retirement thereby encroaching on farming land around cities, contributing to the conversion of an agricultural to a peri-urban landscape.

The second transformation is that of a forested and agricultural landscape to a pastoral landscape, i.e., the investment of remittances in livestock, transforming forested and agricultural land into pasture. The conversion of previously cultivated fields into degraded pasture is also included in this category, with examples from Ecuador, Mexico, Dominican Republic, and Brazil (Jokisch 2002, Lambin et al. 2001).

Another type of landscape influenced by remittances is the transformation of landscapes affected by natural disasters and civil wars through reconstruction. Especially in Central America but also in some African countries, much of the remittances sent back by migrants have helped to rebuild countries after civil war or natural disasters, for example in Honduras, El Salvador, Nicaragua, Lebanon, and Somalia (The Economist 2002). However, reconstructed landscapes are peri-urban, urban, agricultural or pastoral landscapes that were destroyed during natural disasters or wars, and are "restored" to the original type of landscape largely with the help of remittances. As the type of landscape does not change due to the investment of remittances, reconstructed landscapes are not treated here as a third type of remittance landscape.

4.3 Methodological framework

This study investigates land use change as the interface of human-environment interaction dynamics. Gaining a clear understanding of the determinants of land use change requires the use of interdisciplinary research techniques that combine quantitative spatial analysis with qualitative socio-economic analysis. A combination of quantitative and qualitative methods was used to identify the household land use choices behind the observed land use changes. Landsat images from 1990 and 2000 were classified into thematic land use maps and changes between different land use classes were computed. In addition, quantitative (survey) as well as qualitative interviews (in-depth discussions with land managers, government officials, academics and representatives from the private sector) were used to shed light on the underlying driving factors of these changes. This combination of methods allows for the investigation of research questions that are situated at the interface of natural and social science, including this project's central question of whether international migration is a driver of land use change.

Studies on land use and cover change rely strongly on remote sensing (RS) and geographic information systems (GIS) technology, sometimes utilized without field verification. However, research by Fairhead and Leach (1996) on forest changes in West Africa found striking differences in land use changes between findings based on analysis of satellite images and their own findings from intensive anthropological fieldwork. Therefore, a combination of ethnography and remote sensing appears to be the most effective approach, as remotely sensed data can provide focus for research questions and for testing broad-scaled hypotheses (Guyer and Lambin 1993). Furthermore, social science methodologies can be used to improve interpretation of remotely sensed data (Rindfuss and Stern 1988). The results from the remote sensing analysis and from in-depth interviews with local stakeholders complement each other to yield a more holistic perspective.

The spatial representations used in this study are maps generated through analysis of satellite imagery, and written explanations of changes based on knowledge gained through questionnaires and in-depth interviews. Land use and cover changes between 1990 and 2000 were analyzed at the household level and at the municipal level. At the household level, a land use and cover change questionnaire was used with a 30% minimum sample of randomly chosen households owning land in the three rural communities. The same questionnaire was used with three private landowners. The questionnaire focused on total land owned by each household, the area of land use change and the reasons for this change. The methodology description is presented in two parts: the spatial assessment of land use changes and the interview-based assessment of land use changes.

4.3.1 Spatial assessment of land use changes

In order to identify land use changes at the municipal level, Landsat images were classified into land cover classes. For 1990, a digital satellite image (Landsat TM 5 from 7 March 1990) covering the municipality with a surface of 927 km^2 was used. For 2000, a digital satellite image (Landsat TM 4-5 from 21 January 2000) was used. From the digital data, a false color grid composite image (Landsat bands 432) was developed for systematic classification and assessment of land use and land cover change. An MSc geomatics student, Anna Svanberg, under the supervision of Abram Pointet,[16] digitized both images in March 2006. Images were manually digitized at a scale of 1:25'000 and classified into the following land cover classes: Urban, agriculture, dry deciduous forest, pine and pine-oak forest, pasture and shrubland, bare ground and shadows. Data interpretation and analysis were conducted using the software programs Manifold 6.5 and IDRISI 32. All calculations were done using GIS analysis (Manifold 6.5).

Table 4-01 Description of land use classes used for classification of satellite images

Land use	Description
Agriculture	Rainfed and irrigated cultivation including fallow plots
Dry deciduous forest	Forest located mainly between 900 and 1200 masl
Pine and pine-oak forest	Forest located mainly between 1200 and 2800 masl, also includes small patches of tropical montane cloud forest
Pasture and shrubland	Areas used as pasture and areas with shrubs, bushes and some larger isolated trees
Urban	Urban settlement, including small villages and industries
Bare ground and shadows	Exposed rocks and areas on satellite images covered in shadows

During fieldwork between 2002 and 2004, 200 control points were collected by the author with a GPS (Garmin Etrex Summit) and imported as georeferenced vector files with IDRISI 32. For each point, land cover class and altitude was registered. In order to test the accuracy of land cover and land use classification for the image from the year 2000, 53 GPS control points were randomly selected from the 200 collected points and land use classification was controlled. The remaining GPS points were used to improve visual classification. Land use changes between 2000 (classified image) and 2003 (year GPS points were collected) are minimal and should not significantly influence the accuracy assessment. The assessment showed a classification accuracy of 73% with most classification errors occurring in the pasture and dry forest categories.

---

[16] GIS specialist from the Geographical Information Systems Laboratory (LASIG) at EPFL.

The most important land use change in the region is the expansion of agave fields. However, it is virtually impossible to detect agave fields by either remote sensing analysis or visual inspection of satellite images. The first agave fields in the municipality were planted in 1996. At the time the satellite image was taken (2000) the agave seedlings were still relatively small, and as they are planted in rows approximately 4 m apart with virtually no vegetation between them, agave fields have a very similar spectral response pattern to bare ground. Therefore, a GPS was used to register the location of agave fields. The collected data was compared and complemented with data from a BSc study by Flores and Zamora (2003) and from the MSc study related to this PhD research (Portner 2005).

In order to analyze the location of the agave fields with regard to the topography of the municipality, a digital terrain model[17] (DTM) was developed based on elevation contours by Dr. A. Heinimann using ArcInfo.[18] The contour lines (10 m vertical interval) were provided in AutoCAD format by the director of cadastre of the municipality of Autlán and imported as shapefiles into ArcView. Based on the digital terrain model, an analytical hillshade was computed for graphical visualization of the surface where agave fields are located.

Although the contour lines provided by the municipality cover only approximately 90% of the municipality (see **figure 4-02**), no agave fields are located in the areas not covered by the DTM. Average slope for each agave field was computed in ArcInfo with a resolution of 10 meters.

---

[17] A digital terrain model is a map containing elevation information for every point on its surface.
[18] Dr. Andreas Heinimann is a NCCR North-South PhD researcher and GIS specialist at the Center for Environment and Development at the University of Berne.

Figure 4-02 Coverage of contour lines

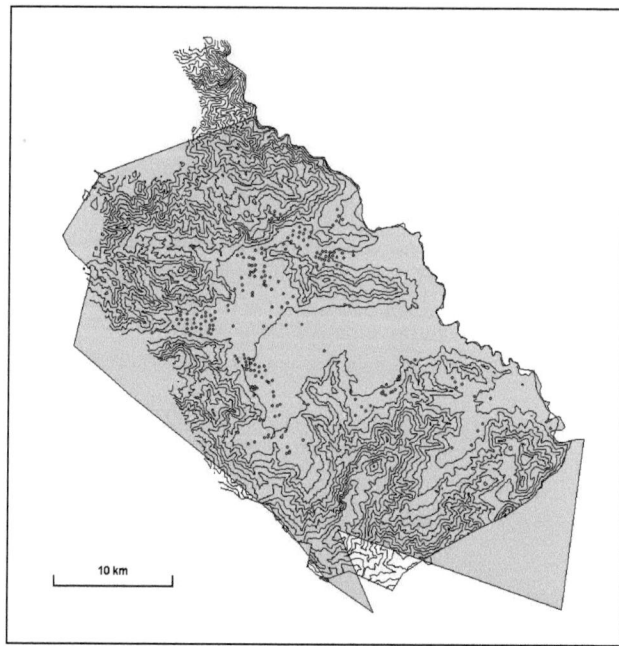

The municipality of Autlán (green area) with contour lines provided by the service of cadastre of Autlán (in black). Agave fields are indicated as (originally) red dots. Vertical interval scale of contourlines on this map is 100m.

A number of municipal, state and federal agencies were contacted for information on land use and land cover at the municipal level for the period 1970-2003. The following data was either provided directly by these institutions or downloaded from their respective websites as Excel spreadsheets. This data was used as comparison with own assessment and the contour lines were used to develop the digital terrain model.

Federal level: Instituto Nacional de Estadísticas, Geografía e Informática (INEGI)
- Topographic map of the municipality of Autlán (1:50'000)
- Thematic land use and land cover map of the municipality of Autlán from 1971 (1:50'000)[19]
- Statistical information on agriculture, vegetation, land use, cattle ranching, population and forestry[20]

---

[19] These maps could not be used for land use change analysis for the following reasons: 1) The maps contain a large number of classification inaccuracies; and 2) The maps were only available as hard copies that would have been difficult to convert into georeferenced data files (size of entire map: 1.5m x 2m).
[20] www.inegi.gob.mx/est/contenidos/espanol/sistemas/cem03/info/jal/m015/c14015_09.xls Cuaderno estadístico municipal Autlán de Navarro, Edición 2003

State level (State of Jalisco, Distrito de riego 05) – Secretaría de Agricultura, Ganadería, Desarrollo Rural, Pesca y Alimentacíon (SAGARPA)
- Land surface (in ha) of pasture, agriculture, forest and urban area for each *ejido* for the year 2000
- Land surface (in ha) planted with *agave azul* for 1996, 2000 and 2004 at municipal level
- Land surface (in ha) planted with maize for 1984, 1990, 1994, 2000, 2004
- Land surface (in ha) with pasture for 1994, 2000, 2004

At municipal level: Municipal government of Autlán (Cadastre)
- Contour lines of the municipality of Autlán (10 m and 100 m vertical interval)
- Various statistics on land use and agricultural production at municipal level

From the Department for Ecology and Natural Resource Management of the University of Guadalajara:
- Landsat image from 21 January 2000 (Landsat TM 4-5)
- Municipal limits of Autlán (ArcView shapefile)
- Various materials, mostly on the Sierra de Manantlán Biosphere Reserve

4.3.2 Interview-based assessment of land use changes

Data collection: Fieldwork took place during 8 months between November 2002 and May 2004. Data collection relied on multiple sources, both quantitative and qualitative. A survey was conducted to collect quantitative data, and interview guides were used to supplement data collection with qualitative data. For the survey two different questionnaires were applied, one focusing on migration and remittances and the other one on land use and land cover changes. The questionnaires were pre-tested twice.

Proportionate stratified random sampling was used. In a first step a household listing was established for each study site; all the houses of the community were schematically mapped and given a number. Empty houses were not included. With the help of key informants and by directly asking each household, the households were stratified into migrant households and non-migrant households. A migrant household is defined as a household that has had a member of the immediate family participating in international migration at some point since 1980. A household is defined as a unit that uses the same kitchen.

The random selection was done by establishing two lists in Microsoft Excel, one listing all migrant households and one listing all non-migrant households. Then the following procedure was used: The household numbers were copied into a column in the spreadsheet. Then, in the adjacent column the function =RAND() was used to assign a random number between 0 and 1 for each cell. Next, both columns were ordinally sorted using the list of random numbers, which effectively rearranges the list in random order. The first 30% of all migrant and non-migrant households were selected for interviews from these sorted lists. If after several attempts no one could be interviewed in a selected household, then the next household in the sorted list was selected.

The survey was augmented with 14 interviews with key persons in the three communities, using an identical interview guide. The key persons were knowledgeable persons in each respective community, such as formal and informal village authorities and elders. These interviews aimed at providing a more holistic picture of land use changes and the driving forces behind them. For the in-depth interviews, the key persons were chosen starting with a local authority (e.g. *comisariado*) and then applying the snowball system. In addition, in-depth semi-structured interviews were conducted with 15 key persons from the municipality, academia and private industry (see list of interviewed persons in *annex 2*).

The land use and migration questionnaires were also applied to three private landowners with properties ranging from 40 to 100 hectares.

A research assistant carried out approximately 30% of the questionnaires for the survey. All interviews with key persons were done in Spanish by the author without using a translator.

The above-described data collection strategy generated a total of 87 interviews using the migration questionnaire, 43 interviews using the land use and land cover questionnaire (not all selected households owned land), 14 interviews with key persons from the communities and 15 in-depth interviews with experts at the municipal and regional level.

Furthermore, two focus group discussions on migration and the investment of remittances were organized. A list of key questions was prepared in advance and used to guide the discussions. The researcher also used participant observation at a meeting of eight municipal presidents discussing sustainable regional development.

In addition, this study integrates the results of an MSc study on the differences in land use strategies of migrant and non-migrant households. The fieldwork of this study was conducted from September to December 2004 by Brigitte Portner of the University of Berne (Portner 2005). This study was also located in the municipality of Autlán in the *ejidos* of Rincón de Luisa and La Laja de Abajo and generated an additional sample of 39 questionnaires and 11 in-depth interviews.

Table 4-02 Overview of interviews conducted

Case study sites	Jalocote	Chiqui-huitlán	Mezqui-tán	TOTAL	*Rincón de Luisa (Portner 2005)*	*La Laja de Abajo (Portner 2005)*
Number of households with migrants	31	23	58	**112**	49	1
Number of households without migrants	26	19	82	**127**	33	21
Total number of households (HHs)	**57**	**42**	**140**	**239**	**82**	**22**
Questionnaires for survey with migrant HHs	11	9	26	**46**	22	1
Questionnaires for survey with non-migrant HHs	7	5	29	**41**	6	10
Total number of questionnaires (number of questionnaires as percentage of total number of HHs)	**18** (32%)	**14** (33%)	**55** (36%)	**87** (**36%**)	**28** (34%)	**11** (50%)
Expert interviews	4	5	5	**14**	6	5
Total questionnaires and interviews in *ejidos*	**22**	**19**	**60**	**101**	**34**	**16**
Expert interviews from private industry/ academia/municipal government				15		
Interviews with private landowners				3		
TOTAL				**119**		
Total including MSc study by Portner (2005)				**169**		

Data analysis: A quantitative analysis of the questionnaires was done using Microsoft Excel. The qualitative data was analyzed for links between migration, remittances and land use and cover change. Data analysis and interpretation of qualitative data followed the procedure described in Creswell (2003). Data from in-depth interviews and the survey regarding the reasons for land use and land cover changes were coded and classified according to six potential driving forces commonly used in land use and cover change research (Agarwal et al. 2002, Geist and Lambin 2002).

An overview of examples of proximate causes for each of the six categories of driving forces is given in **Table 4-03**. Proximate causes usually operate at the local level (individual farms, households or communities). Underlying causes in contrast, may originate from the regional, national or global level (Lambin et al. 2003). The overview is based on a review of proximate causes cited in the literature, which also showed that the distinction between proximate causes and driving forces is far from uniform. Furthermore, a number of proximate causes could be attributed to several underlying driving forces, for instance cattle ranching (economic and cultural) and migration (economic and demographic).

Table 4-03 Overview of proximate causes of driving forces

Driving forces	Proximate causes
Economic	Market prices, investment capital, availability of off-farm employment, cattle ranching, logging, charcoal production, migration
Political and social institutions	Land tenure regime, subsidies, access to credit and markets, corruption, social networks, decentralization, local organizations
Technology	New farming machines, chainsaws, roads, transport, infrastructure, new varieties of seeds
Culturally determined attitudes, beliefs and behavior	Local customs and inheritance laws, traditional shifting cultivation for subsistence, intention to keep farming despite low income, demand for organic products, low concern for nature, absence of stewardship values, attitude towards risk, gendered labor patterns
Demography	Population growth, lack of labor due to out-migration, in-migration leading to population pressure, age and illness as lifecycle features
Environmental	Rainfall, topography, vegetation, soil quality, hydrology, climate change, drought, hurricanes

Source: Own overview based on Bürgi et al. 2004, Lambin et al. 2001, Lambin et al. 2003, Lambin and Geist 2002, Turner et al. 1993.

PART II

5. Differences in land use strategies between migrant and non-migrant households

This chapter is largely based on the MSc study of Brigitte Portner (2005) which is closely linked to the present study. Brigitte Portner conducted fieldwork from September to December 2004 in the municipality of Autlán in the communities of La Laja de Abajo and Rincón de Luisa (see descriptions of these areas in *chapter 2*).

Figure 5-01 Location of study areas *La Laja de Abajo* and *Rincón de Luisa*

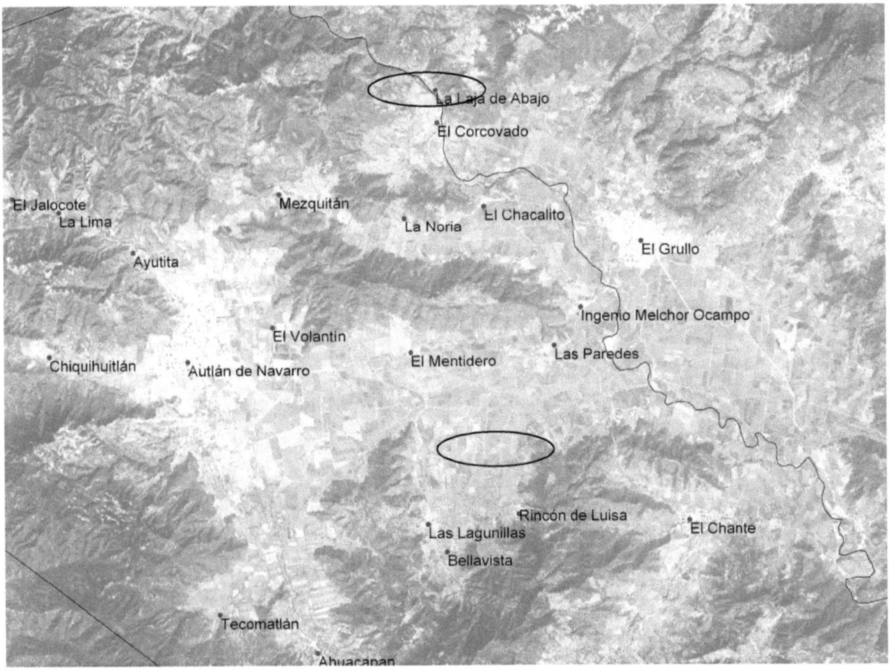

Source: Portner 2005

The objectives of the research project were twofold. One aim was to analyze the dynamics of changes in land use strategies for both migrant and non-migrant households. Second, the changing land use strategies of migrant and non-migrant households were assessed in relation to innovation. In this study, the overall hypothesis was that migrant households have different land use strategies than non-migrant households.

This MSc study addressed two **research questions**:
1. What are the impacts of transnational migration on land use strategies?
2. Are innovations concerning natural resource management leading to remittance landscapes?

The research questions led to the two following **hypotheses**:
1. Land use strategies of migrant and non-migrant households are different.
2. Differences concerning land use strategies of migrant and non-migrant households are a result of greater engagement of the migrant households in the processes of innovation in land use strategies.

The households were analyzed within the sustainable livelihood framework (Chambers 1983, de Haan and Zoomers 2005, Ellis and Biggs 2001, Scoones 1998). Below is an overview of types of capital considered in the livelihood approach:

Human capital: age, education, labor, skills, knowledge, creativity, inventiveness, experience, or good health enabling people to pursue different livelihood strategies.
Natural capital: natural resources useful for livelihood (e.g. land, forest, water).
Physical capital: basic infrastructure and goods (e.g. roads, housing, livestock, machinery).
Financial capital: availability of cash or equivalent means that allows people to adopt different livelihood strategies (e.g. earnings, savings, remittances).
Social capital: social resources upon which people draw in pursuit of their livelihood objectives (e.g. relations of trust, networks, access to institutions).

Each household was rated with regard to these five types of capital (financial, natural, social, physical and human capital) in relation to migration and land use strategies. Innovative land use strategies were identified according to the following criteria:

- The willingness or a plan to change land use to another crop other than maize, agave or sugar cane;
- The willingness or a plan to diversify land use and cultivate several different crops;
- The willingness or a plan to breed cattle for sale and not only for subsistence;
- Differences in cultivation practices such as the reduction of the application of agrochemicals or the willingness to try alternative management practices.

According to these criteria, 27% of households were identified as innovators (8 out of 30 sample households). Six innovators belonged to a migrant household and two to a non-migrant household. All innovators from migrant households come from Rincón de Luisa.

5.1 Livelihood capitals

The following section summarizes the main differences in types of capital between migrant and non-migrant households. It is important to note that there are large differences between the two communities as Rincón de Luisa is significantly wealthier than La Laja de Abajo mainly due to the fact that sugar cane cultivation is possible. Sugar cane yields a relatively high income and is not very labor-intensive. These two factors explain to a large extent why 60% of households in Rincón de Luisa are migrant households. These households have surplus labor since sugar cane is not labor-intensive and also have the financial resources necessary for migration. In contrast, La Laja de Abajo has mainly rainfed land cultivated with maize, and only 8% are migrant households.

Migrant households had a higher level of education than non-migrant households. Taking into account land quality (irrigated or rainfed), migrant households owned more land than non-migrants, and non-migrants were more involved in cattle breeding than migrant households. The explanatory factor however, does not seem to be migration but rather the variations in production systems between the communities. In La Laja de Abajo, which has hardly any migrant households, mainly maize is cultivated. In Rincón de Luisa, mainly sugar cane is cultivated. The difference between these two production systems is that maize cultivation has more potential for pasture than sugar cane. Since pasture availability is often the limiting factor with regard to cattle-raising (Gerritsen and Forster 2001, Young 2002), it seems obvious that the community with the maize production system should own more cattle than the one with the sugar cane production system. In contrast, the study by Young (2002) in the *ejido* of Ahuacapán in the municipality of Autlán found that cattle ownership was correlated with households having members with migration experience. Households with one to five members having migrated are more likely to be involved in cattle-raising than households without migrants or those with more than five migrants in the family. In the study by Portner (2005), only few migrant household invested in cattle; In general, migrant households tended to have small animals such as chickens while non-migrant households raised a larger number of big animals such as cattle. The most important difference between migrant and non-migrant households was with regard to social capital, in particular the involvement in networks. Migrant households were clearly more involved in networks than non-migrant households.

5.2 Case examples

The following chapter presents four case examples to illustrate under what circumstances a household is considered to be innovative. All case examples are from Portner (2005). The examples serve to demonstrate the key issues of migration and agriculture as livelihood strategies. In the first section are two persons with migration experience who are considered to be innovators. These two examples illustrate how land use strategies can be linked to migration. The second section presents two persons without migration experience who are considered to be non-innovators. These examples were selected to illustrate the situation of the least-endowed members of the two *ejidos*. The cases are representative of the two groups (migrant/innovator group and non-migrant/non-innovator group). As physical capital is very similar for all households, it was excluded from the rating and replaced by the categories of "financial capital crops" and "financial capital livestock". The addition of these capital categories increased the visibility of differences between land use strategies. All names of the landowners have been changed.

5.2.1 Innovative migrants

Carlos García

Carlos García is 40 years old. He is the head of a migrant household and belongs to the innovators' group. He is the son of an ejidataria in Rincón de Luisa. He grows sugar cane on 4.7 hectares and maize on 1.3 hectares. He owns some small livestock and has recently bought some cattle. Carlos García also grows alfalfa on the border of the fields for animal feed. *"One year ago we bought the animals. We still use them for home consumption but we hope to sell them in the near future to have an additional income. Today we are learning and trying to breed them and to see if there is a market for the animals. Maize has no value any more and the price for sugar cane has declined"*. Carlos García belongs to the innovator group, as he is trying to establish an additional income from livestock. The relatively high amount of remittances his mother receives is mainly used for food, clothes, and education for his children. This allowed him to invest the income from sugar cane into cattle breeding.

His uncles live in the United States and send USD 1100 per year to his mother, who lives in the same house with him. A joint interview was held with him and his mother as she is the *ejidataria* but he manages the land. Besides the remittances, the PROCAMPO subsidies and the agriculture, they do not have additional income. In total, five persons are living in the household. The interview was conducted by Brigitte Portner on October 21, 2004 in their house in Rincón de Luisa. See **Figure 5-02** for the capital distribution of Carlos García.

After completing his studies at the Bachelor level, Carlos went twice to the US to work. The first time was in 1983, when he stayed for two years. The second time he went from 1990 to 1995. Both times, he worked in construction. He never had the intention to stay for a longer period, as it was clear that he had to take over farm duties in Mexico, since as the youngest son, his mother wanted him to take over the farm. In 1995, he came back and married. He and his wife now have two sons, who attend primary school in Rincón de Luisa. The couple started to be more active in the community when their sons entered primary school, as they are of the opinion that education is not valued enough and teachers do not take their jobs seriously. As no other parent intervened concerning the teachers, they took the initiative to do something about this. They are getting more involved in other community activities as well, having been encouraged to do so by some returned migrants.

The case of Carlos García is rather exceptional for three reasons. First, he started to breed cattle in a period where almost every *ejidatario* in Rincón de Luisa abandoned this activity. Second, he has a university degree, and third, he benefits from the relatively high amount of remittances his mother receives. The fact that he has taken over the responsibility for the land without being an *ejidatario* is quite common in the study area.

It can be assumed that without having the economic security of the income from remittances it would have been unlikely that he would have started to breed cattle. The high human and natural capital resources may have been additional factors that pushed him to change to cattle breeding. The human capital gave him the possibility to obtain and process information: during the interview as well as during informal talks he emphasized the importance of education and said that he frequently goes to Autlán to buy the newspaper. In contrast, the majority of the non-innovators showed much less interest and initiative. Carlos García said that the market value of maize and sugar cane together with the high natural capital he can rely on may have prompted him to diversify his strategies.

Thus, it can be concluded that financial and social remittances have impacts on land use strategies even if they are not visibly or directly invested into land use.

Figure 5-02 Capitals of Carlos García

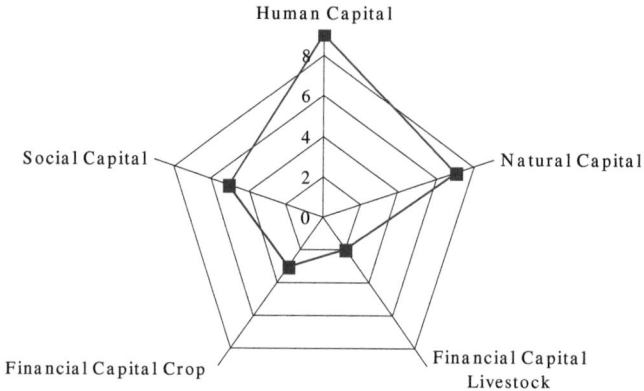

Gustavo Velázquez

Gustavo Velázquez is 60 years old. He is the head of a migrant household and an innovator. He is an *ejidatario* in Rincón de Luisa and cultivates sugar cane on 6.6 hectares. He has over fifty chickens for home consumption. Gustavo Velázquez has close relatives in the US and receives about USD 220 remittances per year. Besides the remittances, PROCAMPO subsidies and agricultural income, their son who lives at home also helps them out. The interview was conducted by Brigitte Portner on December 10, 2004 in their house in Rincón de Luisa. See **Figure 5-03** for the capital distribution of Gustavo Velázquez.

After finishing secondary school in 1959, Gustavo went to Los Angeles, California to work at the same factory as his relatives. At this time, he was still not an *ejidatario* but the son of an *ejidatario*. He says the decision to leave was an easy one: he had no land, his father did not need his labor in the fields, he was not in love with a Mexican woman, and he already knew where to go and where to work. He worked in several factories and lived in Los Angeles for 20 years. Although he spent almost his entire adolescence in Los Angeles, he says he never felt at home in the North because of the vices of the people and his own bad habits. He was whiling away his time in gambling houses drinking and gambling away the money he had earned. *"I didn't live a decent life; I was young and stupid and didn't know the right way."* When he met his future wife they decided to go back to Mexico. At the same time, his father decided to give him the land rights and thus the *ejidatario* rights. This gave him and his wife the opportunity to secure their livelihood with access to land. The money they had saved during their stay in Los Angeles was used to renovate his father's house.

Since 1979, they are back in Mexico and Gustavo works as a farmer on his own land. He abandoned his old way of life to become a good example for his son. His relatives still live in the United States and *"when they have money they send me some because I helped them 27 years ago to go to the North."*

A few years ago, Gustavo fell ill and sold his livestock because *"the doctor said that they are bad for my health."* Nevertheless, he can still cultivate his land by himself, and although it can be painful, he continues to mow weeds using a machete instead of agrochemicals – *"you have to caress the soil otherwise it will not give you anything."* He says that he is not collaborating with other *ejidatarios* nor participating at any meetings as from his point of view, everyone who does, is actually trying to get more power and is therefore lying. Today, because of age and illness, Gustavo Velázquez wants to pass the land to his son, who currently has an off-farm job in Autlán. Gustavo believes that handing over the land would keep his son close to him and his wife. About two years ago, his son saved some money and tried, without success, to cross the border to the United States without the help of relatives. *"It is a pity because if he had told us we would have helped him and now he has to wait until they [the border police] delete his record otherwise they can put him in jail. Well, it is not that bad, this way he is working in Autlán for a good company and he is with us."*

This example shows clearly how important a good and stable network is for successful migration. A typical behavioural pattern of Gustavo and his son is that Gustavo himself returned to Mexico when land rights were given to him and now he tries to keep his son in Mexico by giving him the land. Another point is that Gustavo Velázquez wants to be a good example and live a decent life. For him this includes a careful treatment of natural resources. He has a great awareness of nature and tries to align the needs of his family with those of his land. Application of agrochemicals has been identified as a major concern in terms of the ecological impacts of sugar cane cultivation. In the case of Gustavo Velázquez, it is possible to conclude that his prolonged stay in the United States has led to a more sustainable land use strategy. Or in terms of capital, that a high level of human capital is enhancing the sustainability of his land use strategies. Although one might argue that the reduced application of agrochemicals is also a matter of financial management, the capital distribution of Gustavo Velázquez, his perception and valuation of nature and the availability of remittance income are evidence to the contrary.

The fact that migrants often return to their homeland with different perceptions of nature has been noted by other authors. In one study of the Caribbean islands, Conway and Lorah (1995) found that migrants return to sending areas with a different valuation of ecosystem services and a commitment to preserve the environment and invest in the establishment of local NGOs for the protection of the environment. In another study of the Caribbean, return migrants invested in secure land holdings and in long-term agro-forestry projects (Thomas-Hope 2002, in: Curran 2001).

Figure 5-03 Capitals of Gustavo Velázquez

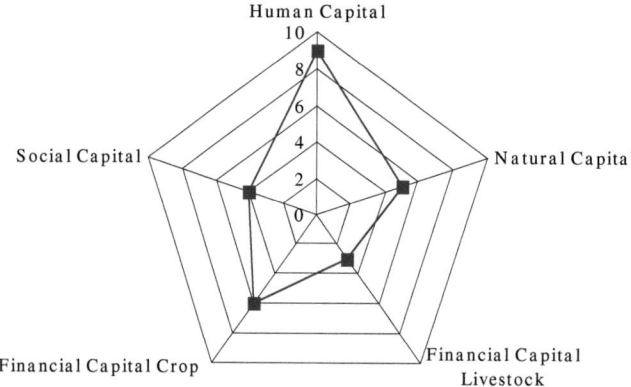

5.2.2 Non-innovative non-migrants

Isidro Santana

Isidro Santana is 70 years old. He is the head of one of the few non-migrant households in Rincón de Luisa. He was identified as a non-innovator in this study. In 1976, two years after his father tranferred the land rights to him, Isidro Santana became an *ejidatario*. He cultivates sugar cane on 3.3 hectares and owns eight chicken. Due to economic problems in 1999, he had to sell the cows and the donkey he owned. Isidro Santana lives with his wife, his sister-in-law, his two children and the two children of his sister-in-law in a small wooden house. The children of his sister-in-law are attending primary school. One of his sons is diabetic and cannot work; the other son is working in the tomato fields in the Autlán-El Grullo valley. Besides the income from sugar cane and his son's earnings, Isidro Santana and his family have no additional income. None of the nuclear family members has been to the US, although they have many distant relatives living in the US or with migration experience. The interview with Isidro Santana was conducted by Brigitte Portner on December 5, 2004 in the front of his house in Rincón de Luisa. See **Figure 5-04** for capital distribution of Isidro Santana.

Isidro Santana has never thought of changing his land use from sugar cane to another crop. *"I have not changed because it is the only crop you earn something by cultivating it. If I had enough money I would cultivate tomatoes or cucumbers. From these crops you get even more."* When he has sufficient financial means he follows the instruction of the sugar cane growers association (CNC): *"when you have money you make more on the field otherwise you wait until the next time you have some money."* When asking him if he had perceived negative impacts from the cultivation of sugar cane he claimed that the burning of sugar cane *"is grilling the animals as well as the trees around the fields."* Even though he perceives this harvest practice as negative he does not want to cut the sugar cane when green, saying *"I have heard of it but I don't know how it works."*

In the evenings he likes to go for a walk in the hills *"to be where no one is and to watch the animals."* Isidro Santana has done this for years and says that because the people of Rincón de Luisa do not go to the hills more animals can be found there than in the hills of the neighbouring *ejido* Bellavista.

Isidro Santana says that the reason he never went to work in the US is because *"I do not like the gringas[1]"*. He once made an attempt to go to the US when his uncle invited him in 1954. In Tijuana another relative was waiting for Isidro Santana and gave him a paper with the password for the 'coyote.'[2] *"The coyote was a huge gringa and I did not understand what she was saying. So, I came back to Rincón de Luisa. I do not know the language and how was I supposed to get along over there?"*

The example of Isidro Santana shows that first, financial means are very important for crops and cultivation practices. He might have changed crops or cultivation practices if he did not have economic constraints. Second, he is a passionate nature observer and notices environmental changes. Third, he probably would change his cultivation practices, in particular the harvest practices, if he received detailed information about alternative practices. Fourth, migration is related to the family cycle: Isidro Santana attempted to cross the border at the age of 20 when he was not married and did not have land rights. Fifth, it illustrates that even with relatives in the US to help arrange the border crossing, successful migration is not assured. The new and unfamiliar situation when Isidro Santana met the female coyote was threatening enough that he preferred to go back to Rincón de Luisa.

---

[1] 'Gringo' or the corresponding female 'gringa' is a popular term used in Mexico for citizens of the United States.
[2] 'Coyote' is the term for people who help migrants to illegally cross the border between Mexico and the United States for payment.

Figure 5-04 Capitals of Isidro Santana

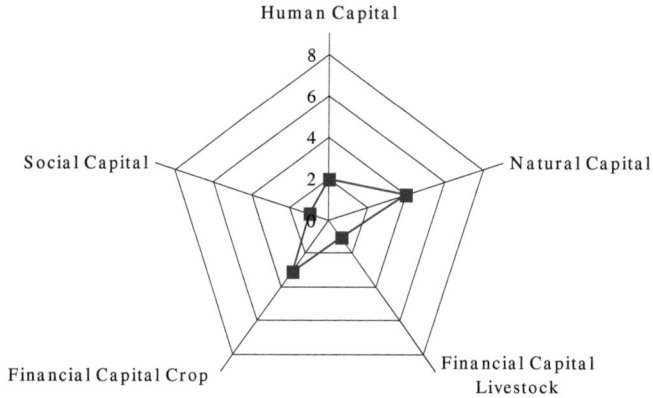

Luz Parrado

Luz Parrado is 45 years old and belongs to the non-migrant group and to the non-innovators. She is an *ejidataria* in La Laja de Abajo but does not cultivate the land herself. Her husband cultivates maize on 3.5 hectares, and they own ten cows as financial security. Four people live in the household: Luz, her husband, their daughter who attends secondary school and their handicapped son. Luz Parrado and her daughter receive subsidies from Oportunidades.[3] Her daughter has to work in the tomato fields in the Autlán-El Grullo valley, as Luz Parrado cannot go to work because she has to stay at home to care for her son. Beside the income from wage labor, the maize income and subsidies from Oportunidades and PROCAMPO, they do not have additional income. No one in their family has relatives in the US or has ever been there. The interview was conducted by Brigitte Portner on November 19, 2004 in front of the family's house in La Laja de Abajo. See **Figure 5-05** for the capital distribution of Luz Parrado.

The family sells about half of the harvest even though Luz said that the remaining half will not be enough for home consumption until the next harvest. They sell the maize to buy fertilizer and to pay back debts. She and her husband are both illiterate. Luz is currently attending classes offered by Oportunidades in order to learn to read and write. She attends the classes because it is one of the conditions for receiving subsidies and she also hopes that then she *"will understand better how things work."*

They have never thought about going to the US. *"We do not know anyone over there and if my husband would go we would not have income from maize anymore."* Luz and her family's

---

[3] Oportunidades is a program of the National Plan for Development (Plan Nacional de Desarrollo) of the Mexican government (2001-2006). The program aims to meet the needs of families in order to enhance their capacity to overcome poverty (Oportunidades 2003: 14-17).

economic constraints are accompanied by environmental ones: their fields have such steep slopes that cultivating maize is the only possibility. In order to have a minimum of economic security they also breed cattle.

The example of Luz Parrado shows that neither migration nor innovation occur among those with very low assets. The main reasons that no one in her family has gone to the US are a lack of financial means, a lack of access to transnational networks, and the hilly location of their fields, which does not allow for the cultivation of a less labor-intensive crop than maize.

Figure 5-05 Capitals of Luz Parrado

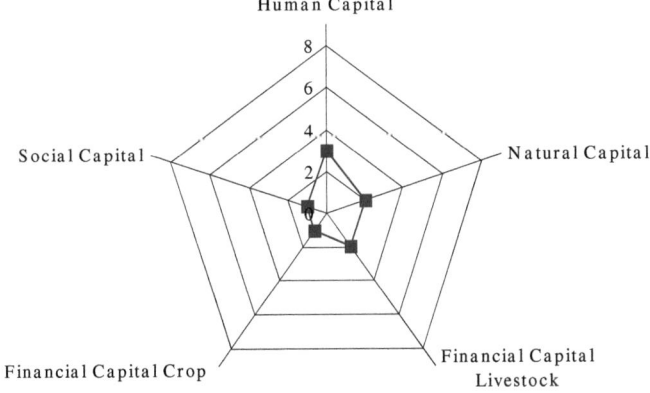

5.3 Conclusions

This study by Portner (2005) confirms the importance of social networks for international migration (Massey and España 1987, Portes and Sensebrenner 1993, Thieme 2006, Thieme and Müller-Böker 2004, Thieme and Wyss 2005). Furthermore, social capital is one of the key factors to access other types of capital. Social capital in the form of access to a transnational network enables people to migrate. In the case of successful migration, this results in higher financial capital due to the inflow of remittances. In turn, this financial capital, allows the household to increase the family's human capital by covering health costs and financing education. Also, it can be invested in agriculture and lead to the adoption of different land use strategies. However, it seems that differences in natural resource management between migrant and non-migrant households are not as pronounced as might be expected (Cassels et al. 2005, McKay 2003). Most migrant households maintain their agricultural practices largely unchanged. One aspect is that cultivating the land is an important cultural activity (Jokisch 2002, Nuijten 2001), and also provides security in case other livelihood strategies fail. Yet in the same study area, Young (2002) found that migrant households tended to buy more cattle and to decrease their agricultural activities.

In this study, differences in land use strategies between migrant and non-migrant families were evaluated with regard to innovation. Out of the households receiving remittances, 22% invested remittances in land or agriculture. These investments did not lead to a visible impact at the landscape level in the sense of an emerging remittance landscape. However, they do affect land use since migrant households have the resources to complete agricultural tasks on time and thus achieve a better harvest. Furthermore, migrant households demonstrate more innovative land use strategies than non-migrant households. 75% of identified innovators belong to migrant households. Innovations include a lower level of agro-chemicals uses, and the purchase of cattle to increase income and reduce risk. However, a key issue is that most migrant households cultivate sugar cane. Sugar cane cultivation is not labor-intensive and provides a profitable income. Therefore, the incentive to innovate does not really exist for sugar cane cultivators as existing production systems are satisfactory.

In conclusion, environmental factors such as topography, soil quality and the availability of irrigation water appear to be more decisive for the land use strategy chosen than whether the household has migration experience.

6. Land use changes 1980-2002

In this chapter, results from the land use change analysis at the municipal level based on Landsat images are first presented and discussed. These results are compared with land use change data from state and federal agencies. Second, the results from the land use change analysis at the community and household levels based on survey data are presented and discussed. The third section examines the environmental effects of the land use change from maize to *agave azul*, focusing on soil erosion. The final section summarizes the main results and draws a number of conclusions concerning observed land use changes and their effects.

6.1 Land use change at municipal level 1990 - 2000

The land use change analysis aimed to detect the changes that occurred during the selected time period and to measure the areal extent of these changes. Initially the aim was to analyze changes between 1980 and 2000. However, the earliest available satellite images with a 28.5 m x 28.5 m resolution – and therefore easily comparable to the satellite images of 2000 with the same resolution – were from 1990 for the study area, which limited the timeframe to a 10-year period.

Three issues regarding land use change analysis based on Landsat images need to be mentioned: 1) The satellite images stem from two different months of the year. In the 2000 image (21 January), the dry season is less advanced than in the 1990 image (7 March). This means that the 2000 image reflects a higher amount of humidity contained in the vegetation, which led to the erroneous classification of 5'874 ha of pasture as dry forest. This can be seen in the original cross-tabulation matrix (see *annex 1*) that indicates changes between individual land use classes between 1990 and 2000. This error was corrected in **Table 6-01** by deducting 5'874 ha from the land use class *dry forest* and adding it to the land use class *pasture* for the year 2000. 2) Municipal limits (ArcView shapefile received from the University of Guadalajara) data is inaccurate and reported a lower total surface area of the municipality than indicated by other sources.[1] 3) The distinction between pasture and agricultural lands is difficult to detect as many agricultural fields are used as pasture so that cattle can feed on crop residues after the harvest. At the time when the satellite pictures were taken, the maize fields had been harvested and were being used as pasture, which made it challenging to distinguish these two land use classes during the classification process. Decision-making during

---

[1] However, indications of surface area of the municipality vary considerably between different sources. Reported totals include 86'364 ha (INEGI 1975), 96'280 ha (Municipality of Autlán 2001) and 92'732 ha (INEGI 1991).

classification was based on a combination of criteria, including a visual interpretation of pixels and first-hand knowledge of the area together with field notes and photographs taken at different times of the year.

Land use changes are presented graphically in **Figure 6-01**. In **Table 6-01**, quantified land use changes are presented in hectares and percentages. Rainfed and irrigated agriculture were not differentiated for two reasons. First, the area of irrigated agriculture remained virtually unchanged between 1990 and 2000. In addition, the additional time required for differentiating these agricultural types for classification would have been considerable.

Figure 6-01 Land use in the municipality of Autlán 1990 and 2000

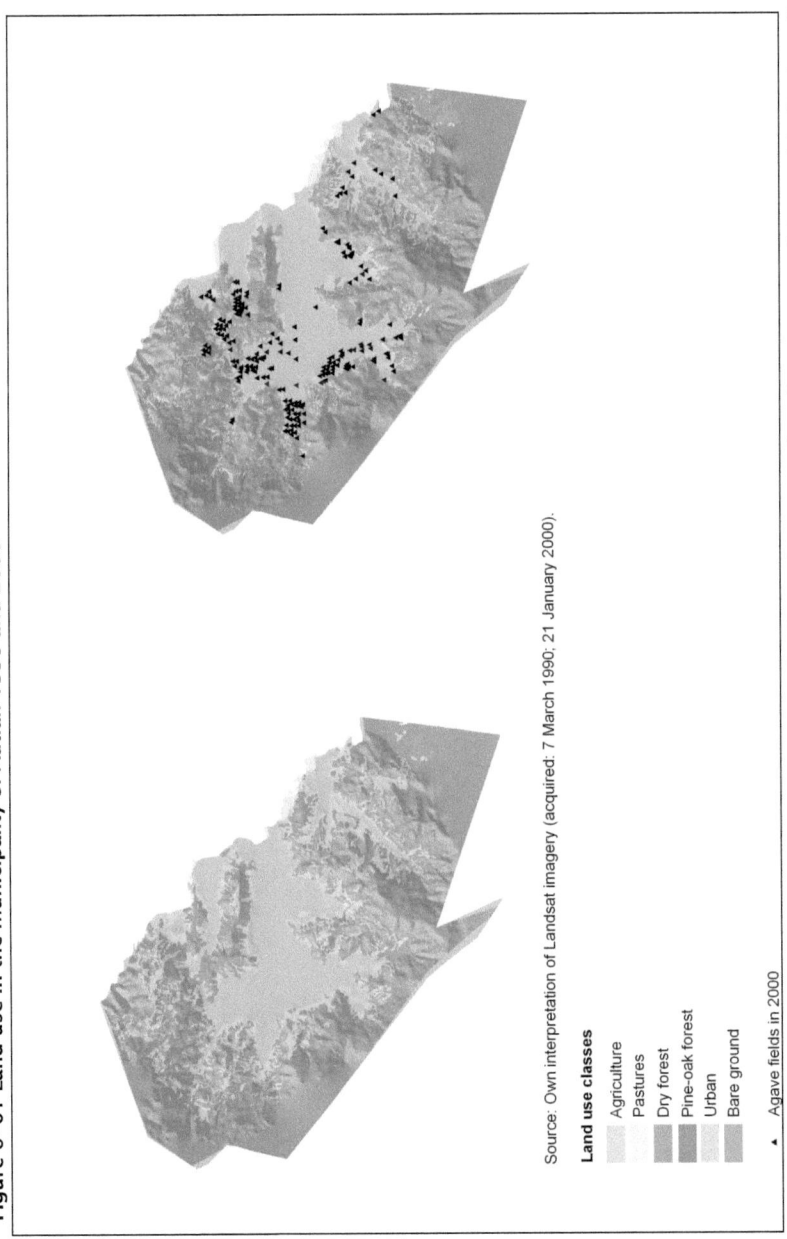

Source: Own interpretation of Landsat imagery (acquired: 7 March 1990; 21 January 2000).

Land use classes
- Agriculture
- Pastures
- Dry forest
- Pine-oak forest
- Urban
- Bare ground

▲ Agave fields in 2000

As mentioned in *chapter 2*, no contour lines were available for the southeastern and northwestern area of the municipality. In reality these "no data" areas consist of hilly terrain.

Table 6-01 Land use change between 1990 and 2000 in the municipality of Autlán

Land use	1990[2] (ha)	2000[3] (ha)	Variation in surface area (ha)	Variation in percent
Agriculture	12'972	13'234	262	2.02
Bare ground	382	167	-216	-56.54
Dry forest[4]	23'625	21'316	-2309	-9.77
Pasture	11'596	13'687	2091	18.03
Pine-oak forest	23'993	24'159	166	0.69
Urban	1'104	1'110	6	0.54
Total	73'673	73'673	0	

Source: Calculated using GIS analysis

Table 6-02 Comparison of land use in 1993 and 2000 at national level in Mexico

Land use	1993 (ha)	2000 (ha)	Variation in surface area (ha)	Variation in percentage
Agriculture	30'198'400	32'803'781	2'605'381	8.63 %
Forest (*bosque*)	34'666'107	32'851'306	-1'814'801	-5.24 %
Dry forest (*selva*)	34'387'491	30'816'633	-3'570'858	-10.38 %
Pasture for cattle (*ganadería*)	27'791'854	31'787'163	3'995'309	14.38 %
Secondary vegetation (*matorrales de zona áridas*)	57'959'607	55'810'305	-2'149'302	-3.71 %
Other	8'886'659	9'820'930	934'271	10.51 %
Total	193'890'118	193'890'118	0	

Source: SEMARNAT 2002

Urban area of the municipality totaled 1110 ha in 2000. It increased by 0.54% between 1990 and 2000. According to Ernesto Medina (2004), the historian of the town of Autlán, the town covered an area of 687 ha in 2000. Thus, the town itself constitutes 62% of all urban land cover, and the

---

[2] Digital false color composite 432 of 7 March 1990.
[3] Digital false color composite 432 of 21 January 2000.
[4] Original results indicated a surface of 23'625 ha dry forest in 1990 and a surface of 27'190 ha in 2000 and a surface of 11'596 ha of pasture in 1990 and a surface of 7'813 ha in 2000 (see cross-tabulation matrix in *annex 1*). This had to be corrected to account for the fact that in the composite of 2000 the dry season was less advanced than for the composite of 1990 which led initially to an erroneous classification of pasture as dry forest.

remaining localities and industrial areas outside the town compromise 38%. The town of Autlán grew 12 times in area between 1966 (urban coverage 57 ha) and 2000 (urban coverage 687 ha). During the same time, urban population roughly doubled (from 27'005 inhabitants in 1960 to 50'846 in 2000) (Medina 2004). Reasons for the steady urban growth include the arrival of immigrants from other states of Mexico to work in the sugar cane, fruit and vegetable industries, and the students who arrived with the opening of a Center of the University of Guadalajara in the town of Autlán in 1994 (Medina 2004).

The decrease of bare ground amounts to 56.54%, which seems very high. However, in absolute numbers, the change is actually very small (an increase of 216 ha in 10 years). Furthermore, the land use class 'bare ground' also includes the area covered in shadows on the image. Therefore, the increase of bare ground is also due to a lower area of shadows on the 2000 image.

Since 1960, the forested area of Mexico has decreased while the area dedicated to cattle and agriculture has increased. Proximate causes responsible for deforestation include expansion of agricultural land and pastures, logging, woodcutting and forest fires (Klooster 2003, Masera et al. 1997, World Bank 1995). In the study area, dry forest decreased by 9.77%,[5] which is very close to the national rate of 10.38% for a similar time period (1993-2000) (**see Table 6-02**). This result also corresponds to the report that deforestation primarily affected dry forest in the state of Jalisco (9% of total surface deforested between 1981-1991) (SEMADES 2006). Several factors affect deforestation in the region: 1) The practice of slash-and-burn agriculture which involves clearing forest to grow subsistence and cash crops; 2) Cattle ranching that requires clearing the forest to plant pasture; 3) Commercial logging; 4) Forest fires (in the region of Autlán, every year 3-8 forest fires destroy between 50 and 350 ha); and 5) Firewood collection and the production of charcoal (Gerritsen 2002, SEMADES 2006).

The category "pine-oak forest" (which includes also other forest types such as fir-oak-pine forests and tropical montane cloud forests) showed an increase of 0.69%, which is in contrast to the national average showing a decrease of 5.24%. This disparity is probably due to the fact that a large part of the area containing oak, pine-oak and fir-oak-pine forest is included in the Sierra de Manantlán Biosphere Reserve, which has stricter regulations concerning forest use than most unprotected forest areas in Mexico. Overall, the deforestation rate (e.g. 5.4% from 1986-1990) in

---

[5] Deforestation is measured as the area deforested between 1990 and 2000, expressed as a percentage of initial forest area in 1990.

the municipality of Autlán is considered moderate by the municipal authorities (Municipality of Autlán 2002).

Pasture increased by 18% between 1990 and 2000. The expansion of cattle ranching is believed to be the primary factor encouraging deforestation in Mexico (Masera et al. 1997). In the municipality of Autlán and in the state of Jalisco, cattle numbers showed a slight increase between 1993 and 1997 but have been stable since then. However, due to a change of the methodology used for establishing cattle statistics by the Mexican National Institute for Statistics, Geography and Data Processing (INEGI), the trends are difficult to assess.

Figure 6-02 Number of cattle in the municipality of Autlán and the state of Jalisco (1993-2002)

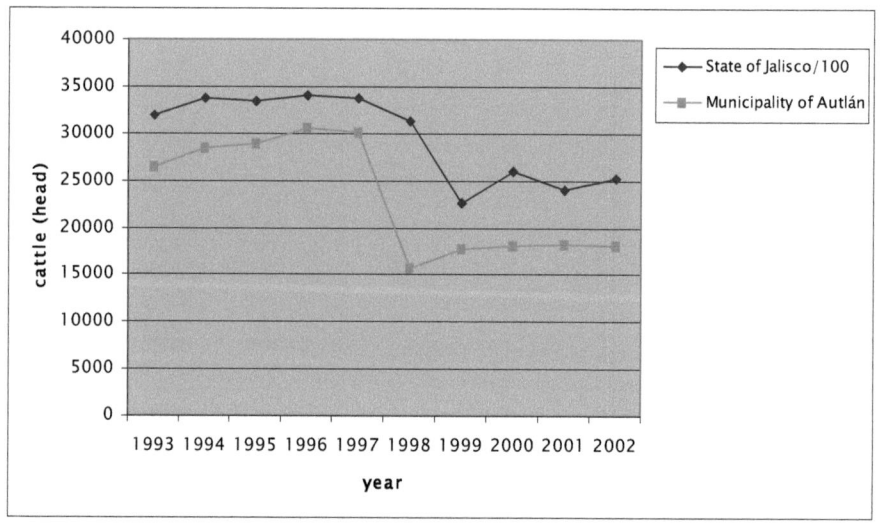

[6] Data source: INEGI 1994, INEGI 1995, INEGI 1996, INEGI 1997, INEGI 1998, INEGI 1999, INEGI 1999, INEGI 2000, INEGI 2001, INEGI 2002.

In two case study sites, the number of cattle was stable due to the difficulty of finding pasture. Even though prices have decreased since 1994, when cheap meat from the US began to arrive under NAFTA, cattle have actually increased in one case study site as well as on the private properties.

Agricultural land comprises 18% of the total area of the municipality, which is very close to the state average; 20% of the area of the state of Jalisco is dedicated to agriculture (SEMADES 2006).

---

[6] The drop in head of cattle between 1997 and 1998 does not represent a real decrease in the number of cattle. It is due to a change of the methodology used for establishing the statistics on cattle by the Mexican National Institute for Statistics, Geography and Data Processing (INEGI).

According to other sources, total agricultural land at the municipal level ranges between 16% (INEGI 1993) and 20% (INEGI 1995) in 1990 and 15% in 2004 (SAGARPA 2006). Farmers are mostly smallholders who raise cattle and cultivate maize, sugar cane, sorghum, chili peppers, and tomatoes. The increase between 1990 and 2000 in the surface area dedicated to agriculture (2%) is well below the national average of 8.63% (**Table 6-02**). However, an important change in rainfed cultivation systems has taken place, with a significant shift from maize cultivation to agave cultivation. The area dedicated to rainfed maize decreased from 5'825 ha in 1996 to 3'789 ha in 2002 (INEGI 2005), while agave increased from 0 ha to 2'254 ha over the same time period. The decrease in the area under rainfed maize (2'036 ha) is very close to the increase in the area dedicated to agave cultivation (2'254 ha), which suggests a direct relationship. This will be discussed in more detail in later sections of this chapter.

Figure 6-03 Area cultivated with various crops and pasture in the municipality of Autlán (1990-2002)

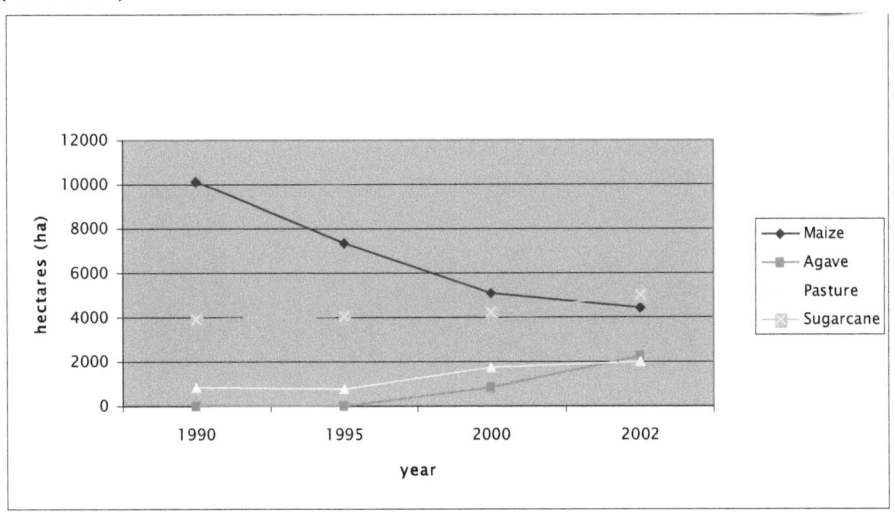

Pasture refers to sown pasture only.
Data source: Statistics published by the Mexican National Institute for Statistics, Geography and Data Processing (INEGI 1991, INEGI 1993, INEGI 1994, INEGI 1995, INEGI 1996, INEGI 1997, INEGI 1998, INEGI 1999, INEGI 2000, INEGI 2001, INEGI 2005) and the Mexican Secretary of Agriculture, Ranching, Rural Development, Fisheries, and Food supply (SAGARPA 2004, SAGARPA 2006).

6.2 Land use changes to *agave azul* 1996 - 2002

The state of Jalisco contributes 20% of national maize production. However, maize production has been decreasing since at least 1990 (**Figure 6-03**). Reasons for this decline are threefold. The first

71

reason is the low profit margin. Until 1990, maize was planted mostly by smallholders, on over one-third of Mexico's agricultural land (Eakin and Appendini 2005). 40% of these producers are subsistence maize farmers (Nadal 2000). In 1994, the North American Free Trade Agreement (NAFTA) between Mexico, the USA and Canada was signed. Under NAFTA the real price for maize in Mexico dropped by 46% between 1993 and 2004 (Eakin and Appendini 2005). At the same time maize imports increased by nearly 200%. Furthermore, the costs for pesticides, herbicides and fertilizer are continually increasing. Therefore, the cultivation of maize is no longer economically profitable (Eakin and Appendini 2005, Nadal 2000) A second reason is the increased variability in rainfall and extreme meteorological events which negatively affect harvests (SEMADES 2006). The third reason is the conversion of agricultural land to urban or industrial use (SEMADES 2006). Although reduction of agricultural land is a problem at the state level, it did not occur in the municipality of Autlán (SEMADES 2006). The cultivation of sorghum and beans has decreased in similar proportions to maize, but the overall cultivated area of maize is much larger (SAGARPA 2006).

There are 136 species of agave (*Agavaceae*), but the only one used to produce tequila is *Avage tequilana* Weber, also called blue agave or *agave azul*. Due to its appearance, agave is often mistaken for a cactus but is actually more closely related to amaryllis and lilies (Mohr 2002).

Agave fields planted around 1998 in Chiquihuitlán. Photo by author (November 2003).

Agave takes 7 to 12 years to mature (Artes de México 1999). Legally, the Norma Oficial Mexicana (NOM) states that tequila is permitted to be grown only within a certain region. Jalisco is the main producer of agave, but it is also grown in the states of Guanajuato, Michoacán, Tamaulipas and Nayarit. Legal tequila must be at least 51% blue agave while premium tequilas are usually 100% blue agave (Mohr 2002). Since the national and international demand for tequila has increased during the past 30 years - tequila sales increased by 1500% from 1975 to 1995 (Mohr 2002) - the tequila companies are constantly in search of new land to expand agave cultivation (Vargas Martin 2003).

Temperature conditions for agave cultivation are a minimum of 3^0 C, an optimum of 26^0 C and a maximum of 47^0 C. A well-draining partially sandy soil is essential, as humid soils tend to propagate disease (Vargas Martin 2003). *Agave azul* should be planted between 800 and 1700 m.a.s.l., with an annual rainfall of about 800-900 mm. The cultivation process of blue agave is complex. Land for blue agave must be cleared and deep-ploughed. The agave is then planted in a straight row, leaving a distance of 2-4 meters between each row and 1 meter between each plant (Vargas Martin 2003). Sowing is done by hand in holes 15 cm deep. Cleaning, pest control, and weeding are carried out annually after the plant is sown. Yields can be between 30 to 200 t/ha, depending on the region (Artes de México 1999). Agave is currently harvested after 7 years. The harvested blue agave plant (core center) is referred to as *piña* (pineapple) because its appearance is very similar to that of a pineapple. The *piñas* provide the raw material for tequila and typically weigh between 15 to 50 kg (González 2002). However, newer varieties can reach a weight of 120 kg (Artes de México 1999). After harvest, it is recommended to let the soil rest for at least one rainy season or to plant a different crop on the same land. However, this rarely occurs due to the large demand for *agave azul*. As a result, fertility problems as well as disease epidemics are a common occurrence in agave cultivation areas (Gomez Garcia 2003).

6.3 Land use changes in agricultural communities and on private properties

For this study, the researcher conducted a survey on land use change history in three agricultural communities and on three private properties. In the agricultural communities, at least 30% of all landowners were included, and the time span considered was 1980 to 2002. Reported results always refer to the sample households included in the survey unless otherwise indicated. In the study region, individually owned rainfed and irrigated agricultural land are referred to as *parcelas*. Land in the hills which is often covered with shrubs or forest is referred to as *cerro* or *sierra*. Results are reported first for agricultural land plots (*parcelas*) and secondly for forested land in the hills (*cerro*).

6.3.1 Land use changes on land plots (*parcelas*)

Of the total land of all case study sites (without the private properties) included in the survey, land use changed on 54% of total agricultural area in the lowland (irrigated and non-irrigated land plots) between 1980 and 2002.

Table 6-03 presents the percentage of landowners changing land use, distinguishing between irrigated and rainfed land. 84% of landowners of rainfed land decided to change land use from 1980-2000. Almost all of these land use changes consisted of switching from maize to agave when the opportunity to cultivate agave or rent out their land to tequila companies presented itself starting in 1996. In contrast, a relatively low percentage of farmers (27%) who own irrigated land decided to change land use during the same period. Those who did change mainly switched from one horticultural crop to another. They consider the economic profits they obtain from cultivating sugar cane, irrigated maize or horticultural crops as good business and have therefore no desire to change.

Table 6-03 Percentage of landowners changing land use, 1980-2000

Land use	Rainfed land	Irrigated land
Change	84%	27%
No change	16%	73%
Total (n=43)	100%	100%

Source: Interviews by author.

As mentioned in section 6.1, at the municipal level the total area cultivated with agave increased from 0 to 2254 ha between 1996 and 2002. This represents 33% of all agricultural rainfed land and 16% of all agricultural land in the municipality. In the two agricultural communities where agave is cultivated, the land dedicated to agave as a percentage of total rainfed land is very high. Within 6 years, it reached 77.2%[7] in Mezquitán, and 99.5%[8] in Chiquihuitlán (Bowen 2004). The main reason for this rapid and extensive land use change of *agave azul* is the fact that cultivating rainfed maize is no longer economically profitable. With an absence of alternatives due to the lack of irrigation water, *agave azul* is a very attractive option.

In the *ejido* El Jalocote no *agave azul* is cultivated. There are three reasons for this absence. First, there are five permanent streams in El Jalocote which provide sufficient water for irrigation agriculture, which earns satisfactory profits. Second, the land plots are small (often not more than 1 ha) and located on very hilly terrain, which is unsuitable for agave cultivation. Furthermore, the soil

---

[7] 201.1 ha out of 260.5 ha
[8] 146.1 ha out of 146.9 ha

quality has been found inappropriate for agave cultivation. In June 2002, an engineer from the company *Agave Azul y Servicios*[9] visited the *ejido* to take soil samples. After the soil was tested at headquarters in Guadalajara, the company decided against renting land for agave cultivation in this area (Vidriales Guzmán 2003).

Table 6-04 Percentage of households (HH) having changed land use on agricultural land plots between 1980-2002

n=39	Percentage of HHs having changed land use between 1980-2002	Land use on remaining agricultural land (ha)
Agricultural communities		
El Jalocote	9% (maize to sorghum)	11 ha of irrigated maize and horticultural crops
Chiquihuitlán	100% (maize to agave)	0 ha
Mezquitán	72% (maize to agave)	81 ha (48 ha irrigated sugar cane, 33 ha rainfed maize)
Private property		
3 private properties	33% (maize to agave)	38 ha of rainfed maize, millet, sorghum and pasture

Source: Interviews by author.

6.3.2 Land use changes in the hills (*cerro*)

In the three community case study sites, on average 52% of the households owning land in the hills deforested part or all of it, mainly in order to establish *coamiles* (see Table 6-05). *Coamiles* are a form of shifting cultivation where the land is first cleared and burned. Then maize is planted and after the harvest, the land is used as pasture where the cattle feed on crop residues. After 3-4 years the land is either left fallow or used only for pasture. 11% of the landowners of Mezquitán plan to deforest in order to cultivate agave in the hills.

For the private properties in the survey, land use changed for 22%[10] of total land (*parcelas* and *cerro* land). Land use change on land plots consisted solely of one private landowner changing 30% of his agricultural land in the lowlands from maize to *agave azul*. In contrast, all landowners deforested part of their land in the hills (9%, 36% and 18% respectively) in order to cultivate agave. However, only one actually proceeded with the agave planting on 18 hectares of his land, while the

---

[9] *Agave Azul y Servicios* (A.A.S.) [renamed to *Azul, Agricultura y Servicios* in 2004] is one of the leading tequila agro-industries. It is an affiliated company that belongs to Casa Cuervo, the second largest tequila production company in Mexico. Tequilas from Casa Cuervo are the most widely sold tequila in the international market and second in Mexico (González 2002).
[10] 47 ha out of a total of 214 ha

second landowner intends to do so in the coming years. The third landowner was so far not able to proceed with the planting because agave seedlings are very costly. He plans to take agave seedlings from his father's plantations and start his own cultivation.

Table 6-05 Percentage of households (HH) having deforested land in the hills between 1980-2002

	Percentage of HHs having deforested land between 1980-2002 (n=41)	Area deforested as a percentage of total area included in sample (ha)
Agricultural communities		
El Jalocote	40% (20% for pasture, 20% for maize)	69%
Chiquihuitlán	50% (50% for maize and pasture)	42%
Mezquitán[11]	66% (55% for maize and pasture, 11% for *agave azul*)	50%
Private land owners		
3 private properties	100% (66% for agave and 34% for pasture)	20%

Source: Interviews by author.

6.4 Assessment of effects of land use change to *agave azul*

Sustainable agricultural production is concerned with the prevention of erosion of topsoil and decline of soil fertility (FAO 2000a, Hurni 2002, Weischet and Caviedes 1993). The ecological sustainability of agave cultivation in the municipality is assessed with regard to the location of the fields in relation to slope inclination. Other factors of soil erosion such as rainfall erodibility, soil erodibility and slope length are not discussed here as a detailed assessment of soil erosion is beyond the scope of this study. Slope is a useful indicator because if sloping land is cultivated without adequate soil protection and sufficient soil conservation measures, there is a risk of accelerated soil erosion. In the short term, cultivation of slopes might lead to a reduction in yield due to the loss of applied fertilizer and fertile topsoil. In the long term, erosion will result in decreased productivity due to a reduction of soil fertility and soil water retention capacity. Soil fertility refers to the ability of the soil to retain and supply nutrients and water in order to allow crops to ideally exploit the climatic resources of a given location (FAO 2000a). Slope gradients are defined according to SOTER[12] for the seven slope range classes used in the land resources database, namely: 0-2% flat, 2-5% gently undulating, 5-8% undulating, 8-15% rolling, 15-30% moderately steep, 30-60% steep, and >60% very steep (FAO 2001).

---

[11] Land in the hills has not yet been distributed to individual landowners in the *ejido* of Mezquitán. However, land distribution is foreseen in the near future. Households in the survey were asked if they would receive land, how many hectares and what they plan to do with it.
[12] SOTER: Soil and Terrain Digital Database, a programme of ISRIC, FAO and UNEP http://www.fao.org/ag/agl/agll/soter.stm

The main soil types where agave fields are located are Phaeozem soils, partially Lithosols and also Cambisols in some areas.

This study is limited to an assessment of the location of agave fields with regard to slope. A detailed analysis of the impacts of agave cultivation on soil quality was not the aim of this project. Ideally, in order to assess the suitability of certain soils for certain crops, soil requirements for the crops should be known. In addition, these requirements must be understood within the context of landforms and other features which do not form a part of the soil but may have a significant influence on the crops that can be cultivated on this soil. The basic internal soil requirements of plants include soil temperature, moisture, aeration, fertility and texture (FAO 2000a). External soil requirements of crops include soil characteristics, slope, micro- and macro-relief, susceptibility to flooding and water-logging (FAO 2000a).

Figure 6-04 Location of agave fields in municipality

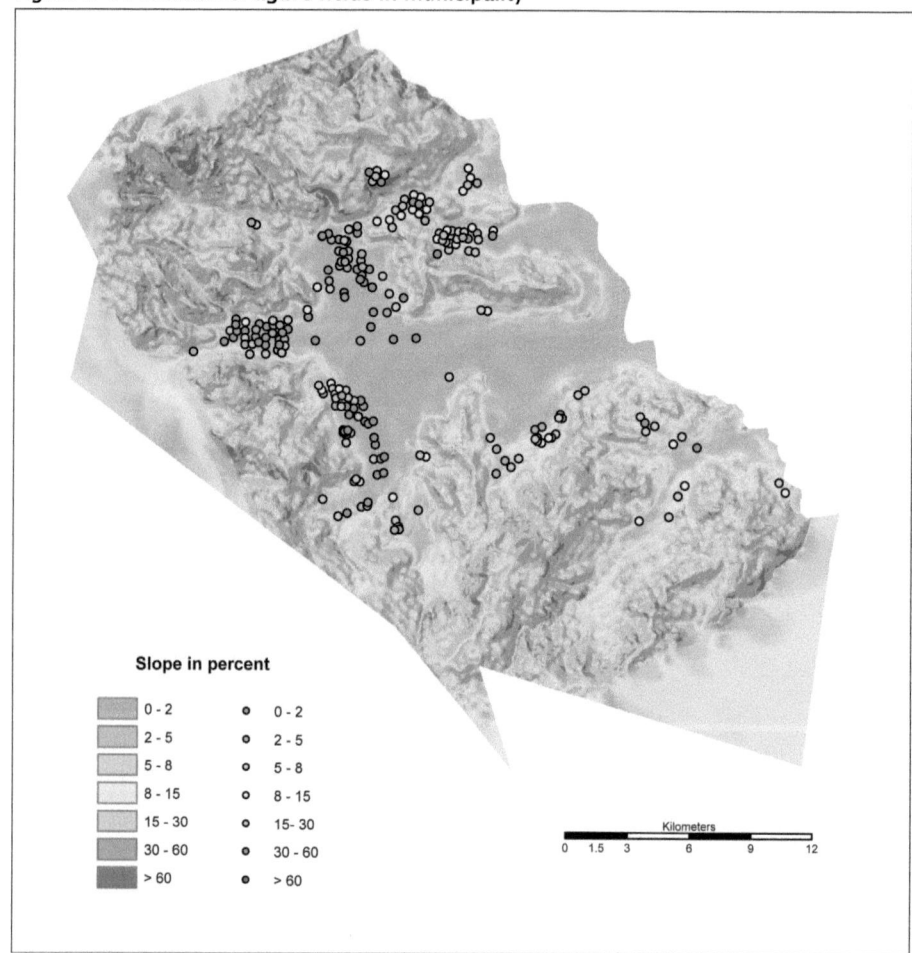

Source: Map created by Dr. A. Heinimann (CDE, University of Berne). Each dot corresponds to an agave field (fields vary in size). The color indicates the slope class of the terrain as well as of the agave fields.

As mentioned in *chapter 2*, no contour lines were available for the southeastern and northwestern area of the municipality. Therefore the slope data for these two areas are not correct. In reality these "no data" areas consist of hilly terrain. No agave fields are located in these areas.

Figure 6-05 Distribution of fields (n = 224) by slope class

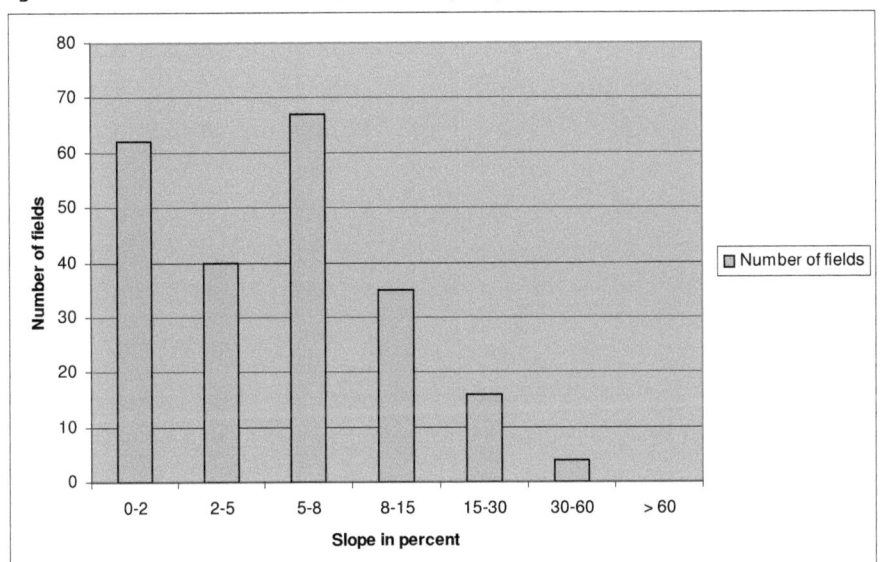

Source: Calculation by ArcInfo (see also **Figure 6-04**)

28% of the agave fields are located in the slope classes 0-2%. Land cultivated on these slope classes requires no conservation measures such as contour cultivation or strips of grass and/or trees along the contours (FAO 2000a). 15% of agave fields are located on slopes between 2-5%. According to FAO (2000a), land with slopes between 2-5% needs no or few conservation measures. However, due to the low vegetation cover during the first four years of the 7-year agave cultivation cycle, conservation measures should already be taken for fields located on slopes over 2%. This would mean that conservation measures are necessary for 72% of all agave fields. Land with a slope greater than 15% should be considered inappropriate for agave cultivation under current management practices. A qualitative assessment of conservation measures indicated a general absence of such measures. On the vast majority of agave fields, the plants are in straight rows perpendicular to the slope. Fields with contour cultivation are a rare exception.

At the state level, agave is cultivated on 55,000 ha. The topographic location of fields has been classified by SEMADES (2006) in the following way: 45% of agave fields are located on flat land (*tierras planas*), 48% on sloping land (*tierras de ladera*), and 7% on stony land (*terrenos pedregosos*). Since "sloping land" does not indicate the gradient of the slopes, a more precise appraisal of the location of agave fields with regard to slope is not possible at the state level and does not allow for a comparison with the situation in the municipality of Autlán.

The cultivation of agave on steep slopes is problematic mainly due to the lack of vegetation between the rows of agave to offer protection against wind and water erosion. Especially during the first 4-5 years, agave plants are still small and provide hardly any ground cover. The soil between the rows of agave (4 m apart) is continually weeded to leave the ground bare. So far no systematic study of soil erosion has been conducted in the study site (Gomez Garcia 2003). However, it is estimated that loss of topsoil during one storm can reach 4 cm (Mejia 2003). The estimation of the degree of soil erosion in the present project (see **Table 6-06**) follows the classification established by FAO (1990).

Table 6-06 Description of various degrees of soil erosion

Degree of erosion	
Slight	Some evidence of loss of surface horizons. Original bio-functions largely intact.
Moderate	Clear evidence of removal or coverage of surface horizons. Original bio-functions partly destroyed.
Severe	Surface horizons completely removed (with subsurface horizons exposed) or covered up by sedimentation of material from upslope. Original bio-functions largely destroyed.
Extreme	Substantial removal of deeper subsurface horizons (badlands). Complete destruction of original bio-functions.

Source: FAO 1990

In the present study, the hazard of soil erosion is estimated based on the distribution of agave fields by slope class (see **Table 6-07**).

Table 6-07 Percentage of fields per slope class and degree of erosion

Slope classes (in percent)	% of total number of fields	Erosion hazard
0–2	28	slight
2–5	17	slight to moderate
5–8	30	moderate
8–15	16	moderate to severe
15–30	7	severe
30–60	2	very severe
> 60	0	extreme
Total	100	

Note: A slope of 100 percent is equivalent to 45 degrees

This means that 28% of agave fields are expected to show signs of slight erosion, 17% signs of slight to moderate erosion, 30% signs of moderate erosion, and 16% signs of moderate to severe erosion, while 9% of fields are likely to show evidence of severe to very severe erosion. This estimation method was also visually tested in the field and found to show an acceptable degree of accuracy. Nonetheless, this analysis is not based on a quantitative survey technique.

On the left: Agave field located on moderately steep land (slope estimated between 15-30 percent) showing signs of very severe soil erosion. Photo by author (November 2003). On the right: visible loss of topsoil during rainy season. Photo by author (August 2002).

6.5 Conclusions

Since 1996, both the municipal government and local residents consider *agave azul* production to be the dominant economic force in the region besides the traditional fruit and vegetable production in the central irrigated valley. However, attention has recently been drawn to the potentially negative impacts of agave cultivation as an economic strategy. In other agave planting regions, a number of studies have reported soil degradation due to the cultivation practices of *agave azul*. However, no coherent land use plan at the municipal level is currently being developed, nor do the tequila companies active in the region collaborate with the municipality in any way (Vargas Martin 2003). The management decisions are made mostly at the household level and with limited alternatives, most households owning rainfed land decide to rent it out to *Agave Azul y Servicios* for cycles of 7 years while those with enough capital opt to become independent agave producers.

Reverse leasing arrangements between tequila companies and landowners transfer the environmental costs of agave cultivation onto the landowner. Tequila companies will move on when harvests decrease or disease begins to affect the plantations. In the study area, a comparison of application rates of pesticides, herbicides and fertilizer showed higher application rates for *agave azul* than for maize, suggesting a higher level of chemical degradation of the soil under agave (Flores and Zamora 2003). If tequila companies decide to leave the area, the landowners will be left with depleted soils on which it will be difficult to revert to growing food crops such as maize, sorghum or beans. The alternative of continuing agave cultivation will only be available to those farmers with sufficient resources for the sizeable initial investment[13] to start planting agave. Furthermore, households starting to cultivate agave on their own account need to have enough other income for the 7 years until the agave harvest. Due to the massive production of agave, the price for agave is very likely to drop. This already occurred between the first planting cycle in 1996 and the second cycle in 2004. In addition, for some of the new contracts, the share of total benefits for the landowners at the end of harvest has been reduced from 5 to 3 percent, while the yearly rent has doubled (from 130 USD/ha/year to USD 260/ha/year).

Rainfed maize cultivation also requires intensive application of fertilizers, herbicides and pesticides, which have led to serious problems of soil degradation (Nadal 2000). One of the differences between agave cultivation and maize cultivation is that for rented land, which constitutes the large majority of agave fields in the region, tequila companies pay for the agricultural inputs and the required fertilizers, pesticides and herbicides are applied in the recommended (high) quantities. In contrast, most households cultivating rainfed maize cannot afford to buy the agricultural inputs that would lead to a better maize harvest. Instead, many farmers apply only small quantities of pesticides and herbicides, which in comparison with agave cultivation causes a lower level of chemical soil degradation. Almost all land owners of rainfed land decided to change their land use to *agave azul* because of the economic benefit to be gained from a crop that does not require irrigation (Bowen 2004, Flores and Zamora 2003). There is only a very small group of landowners who did not have a favorable opinion of *agave azul*. Those very few landowners who decide not to lease out their land for a second cycle of 7 years do so mainly due to the decrease in economic benefits to be obtained from agave and not based on concerns for the environment. During the entire period of fieldwork, only two landowners were encountered who decided not to change maize to agave cultivation out of environmental concerns.

---

[13] Total costs of agave cultivation from initial establishment in the first year (USD 2'500/ha) until harvest amounts to approximately USD 8'000 per hectare (Flores and Zamora 2003).

7. Proximate causes and dynamics of driving forces influencing land use change

This chapter addresses three questions: 1. Which of the six categories of driving forces are the most important ones? 2. What is the relative importance of migration and remittances as proximate causes of land use change? 3. What are the dynamics between the driving forces? In the last section, a number of conclusions are drawn.

7.1 Methods of analysis

In the first section, three case examples for each case study site and three case examples of private properties are analyzed with regard to proximate causes and underlying driving forces of land use changes. The analysis is based on the conceptual framework presented in *chapter 4* and in particular on **Table 4-03** and **Figure 4-01** in *chapter 4.1*. Case examples were chosen based on household income level. For each case study site a low-income,[1] middle-income and high-income[2] household was selected in order to present complementary examples that reflect the overall situation in the case study sites. The first case example of El Jalocote, Doña Mariana Morales, is recounted in more detail than the other case examples in order to give a more complete description of one livelihood and the context in which it is embedded. Even though this case example does not claim to be representative, its aim is to give an idea of the variety of factors potentially affecting livelihoods and land use choices in this area of Mexico.

Land use change is analyzed within the livelihood context of individual households, therefore the time span considered varies between cases. For purposes of comparability, only land use changes that occurred between 1980 and 2004 are included in the summary overview of proximate causes and driving forces. In all case examples, remittances are classified as a proximate cause of economic factors because even though migration itself is a proximate cause of demographic factors, it is rarely the absence of labor due to out-migration that leads to land use changes. Instead, it is the economic situation in the US that allows for the accumulation of capital, which is then invested in Mexico and leads to certain land use changes. Thus, it is more precise to list remittances as the proximate cause of land use change.

---

[1] The level of income is estimated based on indicated sources of income, state of housing, and size and quality of land plots.
[2] In each case study area, three people of the community were independently asked to indicate the three wealthiest households of the community as well as the reasons for their wealth. The case example of the high-income household was chosen from one of these three wealthy families.

7.2 Case examples from El Jalocote

Low-income household: Doña Mariana Morales[3]

Doña Mariana Morales is an *ejidataria* of El Jalocote. She is 82 years old and lives alone in an adobe house without electricity or running water. Her hut can be reached by following a little footpath not far from the main dirt road of the village. It consists of one small room with a bed and another room that serves as a kitchen and living room. She cooks on a handmade stove of earth that she constructed herself. A toilet was constructed about 20 m behind her house in 2002 and is used by several families. Water has to be fetched from a well behind the house. She owns a dog, chickens and a few cats to keep the mice away. Her children bring her maize, onions, chilies, tomatoes and chickpeas. Between 1942 and 1969, Doña Mariana gave birth to 16 children of which 13 survived. Eleven of her children still live in El Jalocote. Her youngest daughter lives nearby and they help each other out on a regular basis. One daughter and a granddaughter who live in Autlán bring her things like coffee, sugar, and soap and sometimes also money. Since 1994, when the PROCAMPO program[4] started, Doña Mariana has received PROCAMPO subsidies for 7 hectares of rainfed land, which amounts to approximately USD 700 per year.

Doña Mariana was born in 1924 in Ayutita and says she had a happy childhood. Her godfather made it possible for her to go to school for eleven months, which was rare at that time, buying her clothes and the necessary school supplies. In 1940, at the age of 16, she married 21-year-old Pedro Morales. Her husband took her to the municipality Villa de Purificacion to live there for one year, because there was work to be found. However, Doña Mariana didn't like it there and never felt at ease, although she did get along well with her father-in-law. She says that she was too young when she married and that she felt lost and disoriented. Moreover, she was in love with someone else but married Pedro Morales because she was afraid of him. He warned her to never see her previous "boyfriend" again, even before they were married. They lived in Cacoma for two years, and she soon became pregnant. Her daughter María was the only one of her 16 children to be born in the town of Autlán with the help of a midwife and her mother-in-law. Her husband helped her during the births of all her other children. She says she taught herself everything, that one knows how to do these things; she doesn't need any medication, which she describes as "chemicals."

---

[3] Name has been changed.
[4] PROCAMPO (*Program for Direct Assistance in Agriculture*) was launched in 1994, to provide transitional assistance to Mexican producers during the implementation of the North American Free Trade Agreement (NAFTA) and the elimination of guaranteed prices for basic staple crops such as maize (www.ers.usda.gov/Briefing/Mexico/Policy.htm). PROCAMPO will have a duration of 16 years (1994-2010).

Doña Mariana is described by others as strong, willful, a hard worker, a well-respected woman and a strict mother. Doña Mariana smoked a pack of cigarettes per day until she was about 70, and even though she only drinks alcohol on special occasions, she is known to be a tough drinker. She is still incredibly fit and goes for long walks in the hills most days. On the way back she collects firewood and usually comes back carrying a large bundle over her shoulder. She describes her husband as a playboy, drinker and a gambler with no sense for business. However, Doña Mariana says she forgave him because he delivered each of her 16 children, which created a strong bond between them.

In 1932 the first land distribution took place in El Jalocote. The land of two haciendas;[5] *"Los Platanos"* and *"Marciano Uribe"* was distributed to the *ejidatarios*. Each *ejidatario* received around 7 hectares. A team of engineers and lawyers arrived from Autlán to organize the land distribution, mandated by the government. Doña Mariana cannot remember what the criteria of land distribution were. There was much disagreement when the land was distributed. Some received better land (location, access to water, soil quality) than others, and everyone wanted more land. Some of the disagreements even led to people being murdered.

In 1943 they arrived in El Jalocote and built their first house. They bought land from an *ejidatario* and constructed their house on this plot. In 1944, a second land distribution took place because new people arrived in El Jalocote and because there was disagreement about the previous land distribution. There was much disagreement also during the second distribution because the people already living in El Jalocote didn't want to lose land due to the arrival of new people. Doña Mariana and her husband received land during the second distribution. They received two land plots, which were assigned in both of their names. They received a rainfed land plot of 5 hectares in the hills, and another irrigated land plot of 2 hectares about 300 m away from their house.

In 1944, the government offered each *ejidatario* 10 head of cattle, which changed some agricultural cultivation from sugar cane to sorghum and maize for cattle fodder. However, not all *ejidatarios* wanted these cattle due to difficulties in finding pasture.[6] Also, many people including Doña Mariana dislike working with cattle as it is difficult work. Furthermore, the cattle are no good for milk. For these reasons, the number of cattle appears to have remained more or less stable, and she only knows of a few migrants in El Jalocote who invested remittances in cattle. Doña Mariana and

---

[5] Large agricultural estates in Mexico that existed until the Revolution in 1910.
[6] Each *ejidatario* has the right to graze 5 head of cattle on the common land. For each additional head of cattle, they must pay 40 pesos (4 USD) per head of cattle for four months. They can sell this right to others, which is often more profitable.

her husband always sold some cattle so that the total number of their herd never exceeded 10. They sold the last head of cattle around 1970. In 1960, they received 10 pigs from one of the deals her husband had made. They kept them for a few years but finally sold them all, as it was not a profitable business. In 1946 they began to cultivate the land plot in the hills. At first they tried wheat but as the land is located at 1700 m.a.s.l. with low and occasionally freezing temperatures, the crop didn't survive the cold period. They also tried potatoes, onions, and millet but none of the crops survived. In the end the only crop that grew was rainfed maize, so they continued to cultivate that for the next 50 years. The plot yields approximately 60 bags of maize.[7]

During the 1950s her husband migrated to the United States to find work. From 1950 to 1956 he picked fruit and worked on cotton plantations in California. The money he earned was used to pay for food and clothes. Doña Mariana is able to identify the exact years her husband was in the US because during this 6-year period none of her children were born. After a few years he left again for six months but this time he spent the money on alcohol and another woman. Doña Mariana thinks migration is only a problem if people leave without sending money back home. If they send money then it is not a problem but a great help. Her description of migrant families is that *"they have money because they have family in the North."*

Doña Mariana has a total of 69 grandchildren of which 17 have migrated to the US, and many more intend to do so once they are old enough or have enough financial capital to allow them to migrate. **Table 7-01** gives an overview of the age of migrants in the Morales family, their destination and their occupation in the US. All of these migrants send remittances to their relatives in El Jalocote, which are an important source of household income. One of her grandsons crossed the border at the age of 17 to work in a restaurant in Oakland, where he earns USD 600 per week. With the money he sent to his parents, they were able to buy food, repay outstanding debts and purchase a truck.

---

[7] Each bag weighs about 80 kg, so the total yield is 4.8 tons of maize for 5 ha. This yield corresponds to the average yield of about 1 ton of maize/ha/year on rainfed land.

Table 7-01 Family tree of Doña Mariana Morales

	9 brothers and sisters					Mariana Morales 1924		Pedro Morales 1919–1991					13 brothers and sisters	
Children	Maria (f, 60) 1942	Manuel (m, 58) 1944	Roberta (f, 57) 1945	Eva (f, 54) 1948	Patricio (m, 52) 1950	Pedro (m, 44) 1958	Antonio (m, 46) 1956	Martín (m, 38) 1964	Ottilia (f, 39) 1963–01	Emilio (m, 36) 1967	Livia (f, 45) 1957	Manuela (f, 42) 1960	Anna (f, 33) 1969
Residency	Jalocote	Jalocote	Autlán	Autlán	Jalocote	Jalocote	Jalocote	Jalocote	Jalocote	Jalocote	Jalocote	Jalocote	Jalocote
Total no of grandchildren	11	12	4	4	6	2	4	4	4	2	8	5	3
Number of grandchildren in migration	2 (m)	2 (m)	1 (m)	1 (m)	1 (m) 2 (f)	1 (f)	–	–	–	–	4 (m) 1 (f)	1 (m)	1 (m)
Age of migrant in 2004 (year of migration)	38 (1989) 41 (2002)	40 (1991) 28 (1996)	38 (1984)	30 (2002)	35 (1986) 30 (1990) 28 (1994)	21 (1999)					28 (1992) 29 (1997) 26 (1999) 30 (2002) 21 (2002)	23 (1998)	17 (2004)
Destiny of migration	California	California Oakland	California Oakland	North Carolina	California	California					California Oregon	Minnesota	California Oakland
Work in US	Plantation	Restaurant	Restaurant	Farm overseer	Restaurant	Restaurant					Restaurant Factory	Factory	Restaurant

Kitchen of Doña Mariana with handmade stove. Photo by author (February 2003).

Doña Mariana with one of her granddaughters. Photo by author (March 2003).

View from El Jalocote towards Autlán. Photo by author (April 2004).

In 1980, the land plot near the village was exchanged for a land plot next to her house (2 ha) irrigated with water from a stream behind the house. The plot next to the house is planted with maize in June, and with chickpea and onions during the second planting season in October. Doña Mariana says that they should never have exchanged these two land plots, because the land plot they originally received had much more irrigation water than the one they exchanged it for. She says it is one example of the bad business deals her husband used to make.

Around 1986, charcoal production started on the common land in the hills of El Jalocote. It is only practiced during the dry season from January to March. People from Colima (a large town located about three hours west from El Jalocote towards the Pacific coast) come to buy charcoal. Many families in El Jalocote work in this sector. A permit to produce charcoal costs 100 pesos (10 USD) for producing one ton of charcoal, and one bag of charcoal can be sold for 40 pesos (4 USD).[8]

In 1991, the husband of Doña Mariana died of cancer at the age of 72 years. The land was from that time on registered only in the name of Doña Mariana.

In 1997, the stream that provided the irrigation water for her land plot next to the house went almost dry. For one year she tried a water pump, but there was not enough water. In the second year she asked the engineer from PROCAMPO for his advice, which was to plant chickpea, but it was too dry and too hot. She tried this for 3 years but now wants to leave the PROCAMPO program because the cultivation of chickpea doesn't work, and there is hardly any harvest, but the engineer insists that she continue. Doña Mariana doesn't know why the engineer insists so much, but she suspects that he must have some personal interest. According to her, the reason that the stream went dry is due to movements of the earth during earthquakes that changed subsurface drainage systems. Some other people in the village think that deforestation in the hills also decreased the water quantity of the stream. The lack of water affects one other family who cultivated maize. Now they use their plot as pasture for their cattle. Other families are not affected by the decrease of water in the stream because they take their water from two other water sources. 24 *ejidatarios* own land located near another stream (Las Iglesias). Water rights are attributed by land plot and not by owner. There are 48 *ejidatarios* in total; the other 24 *ejidatarios* take water from a well. They practice irrigated agriculture and mainly plant vegetables and maize.

In 2000, the engineer from PROCAMPO suggested to Doña Mariana that she rent the land to *Agave, Agricultura y Servicios*, but Doña Mariana doesn't want to. She wants to keep the land so

---

[8] One bag of charcoal weighs about 40 kg.

that her children can cultivate it and she benefits from the products they give to her. With regard to agave she says: *"I have been burned many times [trying out new things and taking risks], I will not do it again."* In addition, if she rents the land to the tequila company she loses control over this plot for 8 years, and she doesn't want that in light of her advanced age.

To the question of what has changed in the *ejido* between her youth and today she mentions several things. *"There was no bus, only a small path, we had to walk to Autlán by foot or with a donkey. All were poor when the sugar cane mill[9] was still working. The hills around El Jalocote were covered in sugar cane. Sugar cane stopped when my daughter María was little [around 1947]. Some people continued to cultivate sugar cane even after the land distribution, but the sugar mill fell into disrepair and the ejido did not maintain it after the haciendas disappeared. And when land was distributed to newcomers they didn't continue sugar cane cultivation. Only very few ejidatarios continued to do so on small land plots. But when the sugar mill fell completely apart, it was abandoned."* The number of cattle in El Jalocote has remained stable. Cattle are sold to the *carnicero* (butcher) or middlemen (traders come to El Jalocote or the people from El Jalocote bring their cattle to them). The price is said to be very low[10] because in 1994 when NAFTA entered into force, cheap meat from the USA flooded the Mexican markets.

The Autlán-El Jalocote road was built around 1987. This was also the time when the people of El Jalocote were able to get work in the towns due to easier transportation options. Bus service started in 1992. There were no schools. It was a difficult life. The poor people didn't have money to pay for school, *"so they stayed stupid."* Her husband didn't go to school at all, so she taught him to read a little bit. Despite her difficulties she thinks she had a very nice life. She thinks that there are no improvements because the government changes one thing and then the next government makes different changes. *"It doesn't matter which government it is, it never gets better, we had so many useless meetings again and again."*

When I first met Doña Mariana she said that forest cover has decreased a lot, but each time I asked after that she said that no, forest cover has always been the same. Her daughters Anna and Eva think that forest cover has decreased a lot due to logging and opening up of *coamiles*[11] and that the

---

[9] The sugar mill was located in one of the two haciendas that owned the land that later constituted the *ejido* of El Jalocote. The hacienda system that began in 1529 was abolished during the Mexican Revolution (1910-1917) and the land was distributed to agricultural communities or as private property to small farmers in the 1930s (Simonian 1995).
[10] The prices paid for a cow of 300 kg is approximately 2100 pesos (210 USD) or 7 pesos/kg for a live cow. People in El Jalocote say that this is a very low price but that it has always tended to be low.
[11] A form of shifting cultivation.

decrease of water is linked to deforestation. But Doña Mariana does not want to agree to this, perhaps because she knows that outsiders such as myself consider deforestation negatively. In previous interviews she complained that the forest had disappeared in order to increase agricultural land. And because there is less forest there is less water and fewer animals and birds. She also says that the soil has weakened due to the application of fertilizers, herbicides and pesticides without letting the land fallow so that it can recover. Her daughter Eva, who lives in Autlán, argues with her family about the negative effects of deforestation, logging and charcoal production. Doña Mariana disagrees, and thinks that the land is there to be used.

The communal use of resources in El Jalocote includes the right to collect unlimited firewood, fruits, (*guayave*, prickly pear, wild tomatoes) and honey. They can also cut old trees, but have to ask the *comisariado* for a permit which is usually granted. In order to establish a *coamil*, the *ejidatarios* have to ask permission of the forest officer in Autlán. The permit is typically given for a small fee without actually visiting the area.[12] The town of Autlán pays one person from El Jalocote to keep the small water stream clean (remove dead leaves and branches and keep the canal functioning), but there is no transfer of funds to the *ejido* of El Jalocote. No logging is allowed in the riparian area since some of the freshwater that flows to Autlán originates there.

Since 2000, one of Doña Mariana's sons put a lot of pressure on her to give him the money from his inheritance. This son pretended that his father wanted him to inherit the land plot in the hills. Doña Mariana says that this son of hers is only after easy money and that he doesn't deserve to inherit anything. She says her husband promised this son some money but not the entire land plot, and she says there were many misunderstandings. *"I have worked hard for this land and will not sell it. It is my right to keep it. Once you lose your land, you spend your money and you're left with nothing."* However, in 2004, after having been pressured for four years almost on a daily basis by her son, Doña Mariana finally decided with much reluctance to sell the land plot of 5 hectares in the hills. A wealthy man from Autlán bought it in 2004 for USD 7'000. She gave USD 3'000 to her son to be rid of him and his demands, but was extremely sad about having had to sell her land. The new owner has planted pasture with the intention to rent out the land. However, due to the high altitude and low temperatures, the pasture has not grown well.

---

[12] According to the law on Ecological Equilibrium and Environmental Protection (LGEEPA), it is not permitted to deforest land on slopes above 15 percent, and a number of seed trees have to be left standing (Martinez 2004). The Law on Ecological Equilibrium and Environmental Protection was approved in 1996. It is the most important Mexican law on environmental issues, delegating responsibility for conservation efforts to the state and municipal level (Gerritsen 2002). Ecological land use planning (Ordenamiento Ecológico del Territorio) is a basic normative instrument and part of the LGEEPA.

Proximate causes: Since 1980, deforestation is driven not only by the establishment of pastures but also by charcoal production and by a logging company active in the area. Land use on Doña Mariana's land plot in the hills changed several times between rainfed crops, motivated by the search for a crop that resists the low temperatures at the high altitude where the land is located. In 2004, land use on this plot changed to pasture when the land was sold and the new owner decided to sow pasture and rent out the land, which is a profitable business. The land use on the irrigated land plot changed in 1997 when the river providing the irrigation water dried up. Even though rainfed crops such as chickpea hardly grew on this plot, the landowner was persuaded to continue cultivation in order to receive PROCAMPO subsidies. The absence of the husband due to migration and the inflow of remittances do not seem to have influenced land use.

Driving forces: Economic, political/institutional, environmental factors.

Middle-income household: Lucio Peredes[13]

Lucio Peredes is 68 years old and an *ejidatario* of El Jalocote who lives with his wife, a daughter and a granddaughter in Autlán. He inherited his land and *ejidal* rights from his father who died in 1952. He used to live with his family in El Jalocote until 1998 when they decided to move to Autlán so his granddaughter Luisa could go to school there. He travels to El Jalocote by bus to work on his fields on almost a daily basis. He is hard-working and well- respected in the *ejido*. People say of him that he is the person who knows best the history of the *ejido*, remembering in detail what has happened in El Jalocote since 1950. He owns 1.25 hectares of irrigated land on which he grows different crops such as tomatoes, maize, millet, cucumber, beans, pumpkins and peppers. His land plot is irrigated from a little stream called *El profundo* which is used to irrigate 12 land plots. The land owners can divert some of the water from the stream to their fields between 6am and 4pm, but the rest of the time the water is used to supply water to the town of Autlán. He rotates crops in order to let the land recover and reduce the danger of diseases, which are more likely to occur if the same crop is planted year after year. He decides which crops to plant based on current market prices. He also owns 2 hectares of land in the hills, which he deforested in order to establish pasture. He now rents it out as pasture as he no longer owns cattle. He registered the 2 ha of pasture with PROCAMPO in 1995 and receives USD 100/ha/year for the land since then. He says that many people opened up *coamiles* in the hills in the early 1990s in order to register it as pasture with PROCAMPO and receive subsidies.

---

[13] Name has been changed.

With regard to the changes in El Jalocote Lucio talks about the important land use change from sugar cane to maize in the 1940s due to the expropriation of the *haciendas*. He says that the *ejido* is developing well due to the new road. He mentions the health center, electricity and the bus service allowing women to go to work in Autlán and bring back money to El Jalocote. He also mentions the big problem of alcoholism in the *ejido*. He thinks that remittances could be invested in a program to combat alcoholism because alcohol is the origin of many other problems such as intrafamilial violence, disputes and poverty. He says it is difficult to say what will happen with regard to the number of cattle; *"Maybe there is more cattle than there used to be because BANRURAL[14] is lending money at very favorable conditions to farmers who want to buy cattle. On the other hand it is difficult to find pasture. Some people will buy and some people will sell."* He also refers to the fact that in the 1980's, *ejidatarios* of El Jalocote rented out pasture to people with cattle from outside the *ejido* due to the high demand for pasture. According to him, about six or seven migrant households have used remittances to buy cattle during the past 15 years. He says that remittances are mostly used to buy food, construct houses and sometimes to buy trucks. Very few buy land with this money as it is hardly sufficient to even cover the cost of basic things. He thinks young people migrate to the US because there is no work in El Jalocote; *"If there was any kind of work here, they wouldn't go over there...there is no work here...they are taking a lot of risks [crossing the border and working illegaly in the US]."* He mentions the fact that most migrants only come back to visit their families and sometimes they take their parents with them and permanently leave Mexico. As negative effects, he mentions family disintegration and the aging of the population in El Jalocote.

He points out that charcoal production which began in 1980 has increased so much that it opens up the forest to an extent that people start growing maize in these areas, gradually converting the forest into agricultural land. Even though people obtain the permits to produce charcoal, these permits are only valid for certain areas in the hills, people also produce charcoal in other areas. He says that there is hardly any control over charcoal production. With regard to the possibility of cultivating *agave azul,* Lucio mentions several reasons why no one is cultivating agave in El Jalocote. First, the land plots are small and located on steep land and the soil is too humid. Second vegetables grow well on irrigated plots so they don't need to plant agave, and more money can be made from vegetables per hectare than from renting the field out for agave. Third, the price of agave is falling,

---

[14] Banco Nacional de Crédito Rural (BANRURAL) gave credit to small rural producers to purchase livestock at advantageous rates. It was replaced by Financiera Rural in 2003. Financiera Rural's mission is to make loans to agricultural producers and rural financial intermediaries. Unlike BANRURAL, Financiera Rural is not a bank, and disperses funds through the branches of affiliated banks.

agave is known to spoil the soil and considerable capital is needed in order to start cultivating agave. For all these reasons he doesn't think changing land use to agave is a good choice.

He has three sons and two daughters. All of them have migrated to the US. Two sons stayed for three years each. The third son left in 1995 at the age of 31 and is still in migration. They work in restaurants and in a greenhouse plantation. His daughter also went to the US at the age of 19 for two years. She returned in 1987 and has lived since then with her daughter Luisa in her parents' house in Autlán. All the migrant children send remittances once a month, and the total amount reaches approximately 1200 USD/year. He and his wife use this money for food, clothes, education for grandchildren and sometimes to pay back debts. None of it is used for agriculture but due to the fact that current expenses are partially covered by remittances, other sources of income are invested in agriculture.

Proximate causes: Land use of the irrigated land of Lucio Peredes is influenced by his desire to maintain soil fertility and the practice of crop rotation, and also determined according to estimated market prices. He has no intention to plant agave because the land is not appropriate for agave and he makes more money producing vegetables. He deforested his land in the hills and has used it as pasture for the past 50 years and receives PROCAMPO subsidies.

Driving forces: Economic, political/institutional, environmental.

High-income household: Luis Espinoza[15]

Luis Espinoza is 56 years old. He and his wife have three children, with two sons in migration in the US, one since 1996 and the other since 2000. With the remittances sent by the older son they financed the migration costs of the second son. One son works as an agricultural laborer and the other works on a construction site. They do not plan to return to Mexico. They left because they were attracted by life in the US and because there was no employment for them in El Jalocote. In contrast, the daughter was never interested in going to the US. Every three months the sons send money to their parents, averaging USD 1200 per year. The parents use this money for food, medical costs, agricultural inputs, car repairs and debt repayment. They also invested it in house construction, and purchased land and cattle. Their house is quite new, brightly painted and one of the few two-story buildings in El Jalocote.

---

[15] Name has been changed.

The family of Luis Espinoza is considered to be among the three wealthiest families of El Jalocote because Luis is hardworking and has a good sense for business. When he was young he spent three years in the US, returning to Mexico in 1976 with enough capital to open a small shop. Their main sources of income are from their shop, from agriculture and cattle ranching, from PROCAMPO subsidies and also the remittances they receive from their two sons.

Luis owns various land plots totaling 3.3 ha, which he has cultivated since 1985. On the irrigated land plot (1 ha) he practices crop rotation to maintain soil fertility. He plants chili, tomatoes, maize, peppers, and pumpkins depending on market price. On the remaining land he cultivates rainfed maize, millet and sorghum.

Since 1990 he has owned 15 hectares of rainfed land in the hills, part of which he purchased with remittances. Before that he rented the land. He deforested in order to establish pasture for his cattle and registered 8 ha of the land with PROCAMPO. He owns 50 head of cattle, the largest herd of cattle in El Jalocote. Most families in El Jalocote own two to three head of cattle, but there are eight families who are significant cattle owners with herds between 10 and 50 head.

Proximate causes: Cattle ranching, subsidies, market prices, remittances.

Driving forces: Economic, political/institutional.

Analysis of driving forces in El Jalocote

The stories of Doña Mariana, Lucio Peredes and Luis Espinoza illustrate several proximate causes of land use changes in El Jalocote. The expropriation of *haciendas* in 1940 followed by the distribution of land caused a large-scale land use change from sugar cane to maize, sorghum and millet. Several factors led to local land use changes. First, the change in land tenure, second, the disintegration of the infrastructure needed for the production of sugar cane and third, the government policy of distributing cattle to *ejidatarios* which created a need for pasture. The need for pasture led to the opening of *coamiles*, which resulted in partial deforestation of the communal lands in the hills. Proximate causes include the change in land tenure and the distribution of cattle, both of which are linked to political/institutional factors.

Between 1980 and 2004, the decrease of irrigation water, maintenance of soil fertility and the location of land plots all constitute proximate causes for environmental factors. In addition, market prices for crops and meat are proximate causes for economic factors that influence land use choices. Finally, agricultural subsidies as a proximate cause for political/institutional factors also drove land use change.

> Overall, the following driving forces and proximate causes underlie land use changes in El Jalocote:
> - **Economic factors**: Cattle ranching, market prices, charcoal production.
> - **Policy and institutional factors**: Subsidies, land tenure.
> - **Environmental factors**: Irrigation water, soil fertility, location of land plots.

7.3 Case examples from Chiquihuitlán

On the left: Boy in Chiquihuitlán. On the right: Agave fields in Chiquihuitlán. Photos by author (March 2004)

Low-income household: Juan Torres[16]

Juan Torres is 61 years old. He has nine children with his wife María who is of indigenous origin and also speaks Nahuatl. His main source of income is his land in the hills and PROCAMPO subsidies. He also owns a herd of cattle but was reluctant to reveal the exact number. Three of his children still live in his house, which is large but in need of repair. His sons work as agricultural workers in the Autlán-El Grullo valley and contribute to the household income. His other children also live in Chiquihuitlán. Some of his sons have constructed houses made of adobe. None of his children have migrated to the US, even though they would like to go, because they do not have the necessary capital to migrate nor access to a network that would facilitate migration. He says that many people of Chiquihuitlán leave for the US in order to find work and earn more money than they could earn from the little work that is available in the region. He says many families do not have enough money or help from the government to survive, so they migrate. In addition, the prices for agricultural products are low, so it is difficult to earn money from agriculture. According to him, migrants from Chiquihuitlán work in agriculture, on construction sites, in restaurants and in factories. In his view migrants do not bring any positive changes to the community. On the contrary, they affect the community negatively because they bring money with which they buy more cattle, which they then graze on the common land without paying. The local authority does not intervene because they are often friends with the migrant families.

While interviewing Juan Torres it is obvious that he has an above-average education as he is very well-informed and uses complex expressions and vocabulary. He reads several newspapers on a

---

[16] Name has been changed.

regular basis. He emphasizes that there are a lot of problems in Chiquihuitlán on several levels. With regard to the environment he mentions the lack of irrigation water, soil contamination, deforestation and decrease of wildlife. He makes reference to a high level of internal conflicts and the lack of employment opportunities. He is very bitter about the injustices that exist with regard to access to common resources due to endemic corruption, which contributes to the poverty of a number of people in the community. He tells anecdotes about government funds that were supposed to finance a canal from Autlán to bring irrigation water to Chiquihuitlán, a project that has been in planning for the past 40 years. According to Juan, the funds disappeared almost entirely into the pockets of the community commissioner and his friends. He says that it is in the interest of the *caciques,* the local bosses, that the canal is never built because if it is, the common land has to be split up on an equal basis accompanied with individual land titles. Since many cattle owners and *caciques* have taken over the use of much of the common land, they oppose the distribution of this land and therefore oppose the canal project. He is also upset because most of the local authorities are interested in making money instead of investing in community development. He claims that there are more wealthy cattle owners who benefit from the government programs for cattle than really poor families under other government programs such as PROGRESA. He says that the laws in Mexico are very good but that they are not applied. With regard to decentralization, he thinks it can be a good thing but only if it is well done, which is not currently the case. In addition, it is wrong that children have to go work for food instead of going to school, and the government should do something about that.

Since 1971, he has owned 40 hectares of land in the hills. He has received PROCAMPO subsidies for 4 hectares of this land since 1994. He is very angry with the local person responsible for registering land for PROCAMPO because this person refused to register the entire 10 hectares of *coamiles* that Juan cultivates. As a friend of the *caciques*, this person can do whatever he pleases. So Juan only receives subsidies for 4 hectares instead of 10.

He considers the rapid expansion of agave azul to be a negative trend for the entire region. Agave drains the soil because of the high applications of fertilizers, pesticides and herbicides and because during seven years no organic matter is produced. He also points out that the company renting the land (*Azul, Agricultura y Servicios*) pays badly. For instance, they only pay for the area actually planted with agave seedlings and not for the path around the field, which means that for a land plot of 3 hectares, 0.5 hectares are lost. However, Chiquihuitlán is dependent on agave due to the lack of irrigation water: *"In El Jalocote they have water, they don't need agave."*

With regard to the land in the hills he observes that there is a continuing trend of deforestation and he thinks there are more cattle now than 20 years ago. He says that it is always the same families who invade common land and use more of the common resources than they are entitled to.

Proximate causes: Cattle ranching, subsidies, corruption, lack of irrigation water.

Driving forces: Economic, political/institutional, environmental.

Middle-income household: Elvira Cardenas[17]

Elvira Cardenas is 34 years old and lives in Chiquihuitlán with her two small children and her husband Jorge. They live in one of the nicest houses in Chiquihuitlán, a recently constructed two-story brick house. The family income is composed of remittances from her husband, income from renting out pasture, PROCAMPO subsidies and salary when her husband is working on agricultural and horticultural fields in the Autlán region. Elvira would also like to open a small shop in Autlán, but does not have the necessary capital and says that it would be difficult as she lives in Chiquihuitlán. Her main worries are related to having enough money and to the fact that her husband drinks too much. She comments on the difficult situation for many women in Chiquihuitlán when their husbands are in migration in the US. They have to make all the decisions and carry the responsibility for the household and the children. She mentions that the biggest problem in Chiquihuitlán is the lack of irrigation water and the inadequate drainage system. Waste water flows through big pipes into a riverbed that is dry most of the year. She says that this leads to much pollution and illness. The access to natural resources is also a problem in her eyes as there is severe deforestation in the hills and therefore less water and fewer animals.

Elvira's husband Jorge practiced circular migration from 1990 to 2002. During this period, he spent 6 months per year picking fruit in California. The reason why he left the first time at 20 years old was to earn enough money to construct a house for his family, which he successfully accomplished. He no longer engages in migration because he considers it is too dangerous for him to risk crossing the border now that he is the father of two small children. However, he mentions that if it was easy to cross the border, the entire family would go. While he was working in the US he sent USD 50 to 100 every two weeks to his wife, which amounts to around USD 1000 per year. The money was used for food, medical costs, education, debt payments, agricultural tools and inputs and above all for house construction. According to Elvira, the situation of her family improved dramatically when

---

[17] Name has been changed.

her husband migrated. She had more money for food and to pay for the doctor and medicine when her children were sick and most importantly, it allowed them to construct their nice house.

The family uses 40 hectares of semi-arid land in the hills as pasture.[18] In 1992, Jorge began to use part of the money he earned in the US to buy fence posts and wire to fence off this land. He estimates that it costs around USD 1500 to fence off 40 hectares. He deforested part of the land to sow pasture grasses but he didn't have to cut many trees as it is semi-arid land dominated by brush vegetation. He also owns a few head of cattle[19] but says that it's a much better business to rent out pasture to other people.[20] Furthermore, he was able to register part of the pasture under the PROCAMPO program and receives subsidies on a yearly basis.

Proximate causes: Cattle ranching, subsidies, remittances.

Driving forces: Economic, political/institutional.

High-income household: Roberto Duran[21]

Roberto Duran is 62 years old. He lives in a big house and keeps a large herd of cattle (around 100 head). Three of his five sons are or have been migrants in the US. Roberto Duran is considered to belong to one of the wealthiest families in Chiquihuitlán because he owns a lot of cattle and has sons in migration in the US: *"They have most of their children in the North and with the money they send they can buy more cattle."* While his sons are in the US he looks after their cattle. They send remittances to their families and also to him. He uses this money for subsistence needs and investment in cattle.

He says that most of the young men who go to the US leave for migration periods of 6-8 months per year and are between 18 and 30 years old: *"Older people have problems getting work in the US, that's why it is mostly the young ones who leave."* He also points out that most migrants leave out of necessity to provide their families with income and to be able to construct a house: *"Sometimes their children are not even born yet and they are not even here anymore."* He says that migrant families that do well are able to improve their house and buy furniture and trucks. With regard to the community, he mentions the lack of irrigation water so that if the rainy season is late or rainfall

---

[18] It was not clear whether they had the right to use this land or whether they had invaded common land.
[19] He didn't want to give the exact number of cattle he owned.
[20] Income from renting out pasture is around USD 4 per head of cattle per 4 months.
[21] Name has been changed.

irregular, they lose the money they spent on buying seeds and agricultural inputs for maize. He mentions also the poor condition of the primary school that needs repair and that the roads should be paved and the electricity network increased. According to him, the remittances sent by migrants do not change the situation of the community, they only improve the individual situation of the migrant families.

He inherited his land from his father in 1964. He owns a plot of 4 hectares in the valley on which he previously cultivated maize, millet, and sorghum in crop rotation in order to maintain soil fertility. In 2002, he decided to rent out this land to the company *Azul, Agricultura y Servicios* (AAS) because he was losing money cultivating maize. He signed a contract with AAS for seven years, during which time he will receive USD 120/year/ha plus 5% of the income from the harvest after seven years. From PROCAMPO he also receives around USD 100/ha/year. He mentions that the price paid for one kilo of agave declined from 15 pesos in 1996 to 3 pesos in 2002. According to Roberto, some families might switch back to maize cultivation again due to the falling price of agave. Many people who have rented out to the *agave azul* company also start working for them as agricultural laborers, or for the large agricultural companies in the Autlán-El Grullo valley. He says that for the large majority of people in Chiquihuitlán, the investment needed to independently start cultivating agave is too high. In addition, seven years is too long to wait for the income from the harvest.

He also owns 6 hectare of land in the hills which he uses as pasture for his cattle and the cattle of his sons. He cultivates *coamil*, using the maize as fodder for the cattle. He had to deforest when he first started to use this land in 1964, and he has received PROCAMPO subsidies since 1994.

Proximate causes: Lack of rainfall and irrigation water, cattle ranching, remittances, low market price for maize and high costs of inputs, subsidies.

Driving forces: Economic, political/institutional, environmental.

Analysis of driving forces in Chiquihuitlán

Deforestation of the land in the hills is driven by an expansion of cattle ranching accompanied by the need for pasture. The investment of remittances in cattle and fencing material also contributes to deforestation in at least one case. The establishment of pasture and *coamil* was partially done with the aim of registering this land for PROCAMPO subsidies. Corruption plays an important role in favoring the invasion (and deforestation) of large areas of common land by the powerful families in Chiquihuitlán. Furthermore, Chiquihuitlán is a *comunidad agraria* with no maps of land plots, and the common land has not been split up among individual households. The absence of individual land titles combined with a certain level of clientelism has favored the deforestation of common land. The majority of the *comunidad agraria* has to vote in favor of participating in the PROCEDE program which would regularize land titles, and this has not yet happened.

Proximate causes for the large scale land use change from rainfed maize to agave cultivation include the lack of irrigation water, the risk of failed harvests due to an increasing variability in rainfall, and the high cost of agricultural inputs combined with the low market price of maize. Furthermore, renting out the land to tequila companies has two additional advantages: 1) The landowner continues to receive PROCAMPO subsidies and 2) He does not have to farm his own land and is free to find off-farm employment, thereby diversifying his sources of income.

Overall, the following driving forces and proximate causes underlie land use changes in Chiquihuitlán:

- **Economic factors**: Cattle ranching, market prices, remittances.
- **Policy and institutional factors**: Subsidies, land tenure, corruption.
- **Environmental factors**: Irrigation water, variability of rainfall.

7.4 Case examples from Mezquitán

A street in the village of Mezquitán. Photo by author (April 2004).

Irrigated sugar cane field in El Volantín belonging to an *ejidatario* of Mezquitán. Photo by author (April 2004).

Low-income household: Francisco Arechiga[22]

Francisco Arechiga is 76 years old. He is an *ejidatario* of Mezquitán but lives in El Volantín where his land is located. He has two children. One of his sons migrated to the US for one year in 1986 in order to earn enough money to construct his own house in Mexico. However, he earned too little and spent most of it on alcohol. During his time in the US, he worked on construction sites but didn't manage to send any money to his parents. Now he is married and does not have the capital to migrate again, and also considers it too dangerous. His father thinks that everyone would like to go and work in the US but it is dangerous and expensive to cross the border.

Francisco owns 9 hectares of rainfed land which he inherited from his father who originally received the land during land distribution in 1942. His income is from crop cultivation, cattle ranching and PROCAMPO subsidies. He plants maize, beans, pepper and millet, changing between the crops. In 1999, he decided to rent 4 hectares of his land to a private owner who planted *agave azul*. He continues to cultivate the remaining 5 hectares as before. The reason why he decided to rent his land was mainly the incentive of receiving a truck in exchange. He also needed the additional income from renting the land to cover his living expenses and agricultural inputs: *"I need machinery to cultivate and I need to meet other expenses and I'm not always able to pay."* He rented out his land for seven years, receiving USD 1000 per year for the four hectares which will amount to a total of USD 7'000. As the truck he received is worth USD 3'800 the private owner still owes him USD 3'200. He says that after this contract runs out he doesn't intend to renew it because

---

[22] Name has been changed.

agave azul is ruining the soil. He adds that the company *Azul, Agricultura y Servicios* does not want to continue renting some of the land because it has become degraded from the high levels of fertilizers, herbicides and pesticides involved in *agave azul* cultivation. He says he will have to leave the land fallow for at least a year, adding organic fertilizer so that it becames usable again. He also keeps a small herd of cattle which he would like to increase but is unable to do so due to pasture limitations: *"I could make more money if I had more cattle but the problem is finding enough pasture for them. In addition, I have to pay 35 pesos [USD 3.5] per hour for water from the well."* Once the common land of the *ejido* of Mezquitán is distributed, he intends to use his share as pasture. He will fence it off and buy goats to graze there. According to Francisco, there are less trees now than 20 years ago and this is why it is getting hotter all the time.

Proximate causes: Market prices, high costs of inputs, access to capital, change in land tenure, soil fertility.

Driving forces: Economic, political/institutional, environmental.

Middle-income household: Teresa Fregoso[23]

Teresa Fregoso is 55 years old. She has been an *ejidataria* of Mezquitán since 1973 and owns 8 hectares of rainfed land. She and her husband have four children, with one son who still lives with them. One son worked in the US on construction sites from 1990 to 1996 because he couldn't find work in Mexico. He left with the idea of saving enough money to build a house upon his return to Mexico. According to his mother he spent most of it on alcohol. One of her daughters left for the US in 2002 and worked as a nurse for elderly people. She got married in the US and will probably not return to Mexico. Both of these children regularly send remittances averaging USD 4800 per year. Her daughter has recently stopped sending money as she now has her own family to look after. Her mother is happy for her because her daughter has a good life in the US and is earning money. Nonetheless she is sad that she never sees her or her grandchildren. She uses the money from remittances to buy food, to finance school expenses for the younger children and to pay back debts. Her other two children are not interested in going to the US as they have found work in the region.

Teresa used one room of her apartment to establish a little shop selling soap, canned food, toys, music tapes, etc. In addition to the income she earns from the shop, another daughter who works as

---

[23] Name has been changed.

a lawyer in Guadalajara also contributes to cover household expenses. The third source of income is from cultivating their land and the PROCAMPO subsidies they receive. She and her husband used to plant maize, millet, and sorghum on their 8 hectares of rainfed land. In 1996 they decided to rent out 6 hectares to the company *Azul, Agricultura y Servicios* (AAS). In 1998, they started to cultivate agave on their own on the remaining 2 hectares. Remittances received from their children together with the income from their shop provided them with enough capital to finance the establishment of the agave plantation. The main reason for this land use change is that both she and her husband suffer from diabetes and are no longer able to work in the fields. Furthermore, cultivating rainfed maize was no longer economically profitable, as there is hardly any harvest due to low and irregular rainfall. By renting out the land to AAS they received USD 120/ha/year and 5% of the total income from harvest. They have signed a second contract for seven years with AAS for 2003-2009. The terms of the contract have changed to USD 240/ha/year and 5% of the total income from harvest. The yearly rent has been doubled to compensate for the fact that the price for agave has decreased from USD 1.5/kg to USD 0.20/kg. Teresa says that a few people in Mezquitán hesitate to rent out their land again as there are many complaints against AAS. They are said to pay little, pay late and when the price of agave fell incredibly there was no guaranteed minimum price. *"However, without irrigation water we do not really have another choice but to rent out again. Because even if the price falls, agave still pays better than maize and is less work and a guaranteed income."* Teresa Fregoso and her husband are part of the association of independent agave cultivators, which meets once a month. At regular intervals agricultural engineers come to advise them on management practices for their agave plantations. She says that the association is trying to solve the problem of finding a buyer for the agave harvest. In addition to agave production, once the common land is distributed to the individual *ejidatarios*, she plans to fence off her 9 hectares and establish *coamiles*.

Proximate causes: Market prices, high costs of inputs, low and variable rainfall, remittances, change in land tenure, labor availability.

Driving forces: Economic, political/institutional, demographic, environmental.

High-income household: Rafael Cortes[24]

Rafael Cortes is 54 years old, an *ejidatario* of Mezquitán and owner of 5 hectares of rainfed land. He has four children, of which two still live at home. He works for the *Comisión Federal de Electricidad* (CFE) and managed to get jobs there for two of his sons. His third son left at the age of

---

[24] Name has been changed.

18 for the US, as he was not employed by the CFE. He works on construction sites and every two months sends USD 100 to his mother, which she uses for food and to cover medical expenses. Rafael's daughter works in Mezquitán.

Rafael received his land from his father in 1992. Until 1998, he cultivated rainfed maize and then decided to plant agave on his own because with limited rain, there is no profit to be made with maize. Despite having to wait for seven years until the agave can be harvested, he says it is worth it as it pays very well. As long as there is no irrigation water in Mezquitán he will plant agave. If irrigation becomes available he will switch to sugar cane.

The most important source of income is the salary Rafael earns working for the *Comisión Federal de Electricidad*. This job provides his family with a secure income, which covers living costs and also provides him with capital to invest in agave cultivation. He works a lot, dedicating most of his spare time to agave cultivation and to the association of independent agave producers. The association is considering selling their agave harvest to a small tequila factory in the nearby town of El Grullo or to a production facility in Tonaya. He points out that the expansion of agave has provided employment for a lot of people in the region, especially the landless. He says that when he receives his share of the common land (9 ha), he will deforest, fence it off, and plant *coamiles*.

Proximate causes: Market prices, low and irregular rainfall, lack of irrigation water, availability of investment capital through secure off-farm job, change in land tenure.

Driving forces: Economic, political/institutional, environmental.

Analysis of driving forces in Mezquitán

The lack of irrigation water combined with low and variable rainfall is the proximate cause for environmental factors and is partially responsible for the extensive land use change from rainfed maize to *agave azul*. The other proximate causes are economic factors, namely the low market price and high costs of inputs for maize, which render maize cultivation economically unprofitable. Quite a few landowners in Mezquitán have several sources of income, either remittances or well-paying off-farm jobs, which are also proximate causes of economic factors. This income constitutes an investment capital which is often used to start planting agave as an independent cultivator.

The fact that agave is eligible for PROCAMPO subsidies is an essential factor in the decision-making process of the landowners. PROCAMPO subsidies constitute a secure source of income for all landowners, no matter whether they decide to rent out their land or to start cultivating agave themselves.

The proximate cause of land tenure is included because all interviewed *ejidatarios* of Mezquitán indicated that once the common lands are split up through the PROCEDE program, they will deforest the land plot that they receive.

Mezquitán presents a unique situation with regard to proximate causes of demographic factors. It is the only case study area where it was frequently mentioned that the opportunity to rent out land to tequila companies was a perfect solution for the elderly and sick, especially elderly widows as they can no longer farm their land alone and often do not have the necessary resources to hire labor to work their fields. The lack of labor is included as a proximate cause of a demographic factor.

Overall, the following driving forces and proximate causes underlie land use changes in Mezquitán:

- **Economic factors**: Market prices, remittances, investment capital.
- **Policy and institutional factors**: Subsidies, land tenure.
- **Environmental factors**: Irrigation water, variability of rainfall.
- **Demographic factors**: Labor availability.

7.5 Case examples of private property owners

High-income household: Rancho Milenio - Elias Vargas[25]

Area where private property *Rancho Milenio* is located. Photos by author. Photo on the left was taken during the dry season (April 2004) while photo on the right was taken during the rainy season (August 2002).

The ranch of Elias Vargas is located in the north central area of the municipality of Autlán. It can be accessed on a well-maintained large dirt road within 20 minutes from the main road linking Guadalajara to the Pacific Coast. Elias Vargas is 56 years old. He has never migrated to the United States. In 1998, the ranch was for sale for USD 450'000 but he was able to negotiate the final price down to USD 280'000. He is quite wealthy and bought the ranch as an investment. He is the owner of a telecom company in Guadalajara where he lives and works. He can reach his ranch within three hours. According to one of his staff, the ranch has changed owner three times since 1980. There are several buildings on his land, including housing quarters for employees, some for storing machinery and tools, and the most recent a weekend house which he built for himself. He employs 10 people to work on the ranch on a permanent basis; they are all natives of the region. During the work-intensive periods in August and September he employs additional workers.

Elias Vargas is also an *ejidatario*. He was the first *ejidatario* in the area to independently plant agave on 8 hectares. The profits he made from his *ejidal* land allowed him to plant more agave on the private property he bought. He is now renting even more land from neighboring farmers on which he also plants agave.

The ranch comprises 70 hectares in the hills plus 30 hectares of agricultural land. In addition, as an *ejidatario* he has the right to use 9 ha of common land in the hills. Before he bought the land,

---

[25] Name of landowner and ranch have been changed.

maize, sorghum, millet and some horticultural crops were cultivated. In 1998, he changed the land use on 16 hectares from rainfed maize and sorghum to agave. As he owns 60 head of cattle and a few horses, he continues to cultivate rainfed maize, millet, and sorghum on the remaining land for fodder. There are also a number of fruit trees and a large vegetable garden on his land. In 2000, he deforested 7 hectares in the hills to use as pasture. However, he is now also considering to plant agave there.

Elias is one of the founders of the association of independent agave cultivators, which includes about 170 independent agave cultivators from throughout the region. Their objectives are to obtain a fair price for agave at the time of harvest and become more independent from the company *Azul, Agricultura y Servicios*. The association also aims to register all producers and the amount of agave that they have planted to verify how much agave is grown in the region. One of the major problems of the independent producers is to find a buyer at the time of harvest. However, *"the real money can only be made by those who cultivate the agave themselves. The other ones who rent out their land get exploited by intermediary dealers who buy the agave and then sell for twice that price to tequila companies."*

The manager of Rancho Milenio says: *"Agave helps to improve the livelihood of the people here, especially the widows."* One of the permanent staff working on the ranch explains: *"If land is rented out to Azul, Agricultura y Servicios, the annual rent amounts to 1'300[26] pesos/ha which is more or less the income earned from one ton of maize produced on the same area of land. The advantage of agave consists in the fact that if the land is rented for agave, no labor has to be invested and no financial resources for fertilizer or pesticides."* If maize is cultivated by hand, the production costs (approx. USD 270/ha/year) are twice as high as the value of the harvest (approx. USD 130) (Flores and Zamora 2003). In contrast, for cultivation of *agave azul*, the total cost of fertilizer, pesticides and labor amounts to approximately USD 237/ha/year. [27] Together with the costs of initial establishment in the first year (USD 2'500/ha), total costs of agave cultivation until harvest amount to approximately USD 8'000/ha. In 2001, cultivators based their calculations on the current price paid for agave, which was USD 1.10/kg with an average harvest of 100 tons/ha. The income from the harvest of one hectare of agave was estimated at USD 118'625 including the

---

[26] Approximately USD 130.
[27] Approximately 250-350 kg of fertilizers are applied to one hectare of agave per year. One bag of 50 kg costs around 120 pesos (USD 12). Cost of fertilizer/ha/year: <u>USD 72</u>. Pesticide application is 1-6 l/ha depending on the product. Expensive products cost around 400 pesos (USD 40) per litre. Total cost of pesticides/ha/year: <u>USD 120</u>. Average labor costs/ha/year: <u>USD 45</u>. **Total costs/ha/year: USD 237** (Flores and Zamora 2003).

benefits made from selling seedlings,[28] for a net income of USD 110'000 (Flores and Zamora 2003). However, prices decreased to around USD 0.30/kg at the actual time of the first agave harvest in 2003 due to the intentional strategy of the tequila companies to rapidly expand agave production. Thus, net income from the first agave harvest was around USD 40'000/ha (Cruz Mercado 2004), which even at these lower prices is still very profitable. According to calculations by the University of Guadalajara, even if the price falls dramatically, e.g. USD 0.10/kg, the independent cultivation of agave is still likely to be the most economically viable land use choice (Bowen 2004).

Management decisions regarding the agave fields on Rancho Milenio are made by an engineer who manages the agave fields of several owners in the region. He decides the timing and amount of fertilizer and pesticide application on a regular basis. Several agave cultivators together cover the costs for his consulting services.

A worker on the farm commented that the advantage of agave is that it provides more employment. Whereas maize just requires labor during planting, agave needs to be weeded continually and therefore provides more work to more people. He does think the wages should be higher than USD 45/week: *"Farm work is harsh, above all while earning so little."* For this reason, two of the workers on Rancho Milenio, Paco and Roberto, migrated to the US. Paco went to work in a restaurant in Texas for three months in 1984 but he didn't like it and came back. Roberto tried to cross the border in 2001 but was caught and sent back. They think that most migrants leave with the idea of earning enough money in the US to construct their own house in the region of Autlán. But once they are in the US, this plan is abandoned and temporary migrants become permanent migrants.

Proximate causes: Cattle ranching, market prices, investment capital, irrigation water.

Driving forces: Economic, environmental.

---

[28] In the third year of cultivation, some producers were able to sell agave seedlings (*hijuelos*) for USD 0.90/seedling from their plantations, making a total benefit of USD 4500/ha. Seedlings can be taken off the plantations from the third to the sixth year of cultivation.

High-income household: Rancho la Luna - Oscar Sepulveda[29]

Oscar Sepulveda is 58 years old. He has four children, none of which ever migrated to the US because they all have work in the region and prefer to live here. His son adds, *"We all have to work, whether it is here or in the US."* The Sepulveda family members derive their main income from a butcher shop they own in the town of Autlán. They also own a large private property, Rancho la Luna, about an hour from town. The Sepulvedas are a very wealthy family, and have owned their land for several generations. Their property includes 20 hectares of rainfed land and 50 hectares of *cerro* land. In 2000, they converted 18 hectares of *cerro* land into agave plantations. As it was semi-arid land, there was limited need to deforest. The remaining land is all used as pasture for their large herd of cattle (approximately 80 head). No maize is planted. The change to agave seemed to be a good investment as the price of agave was high at that time. Furthermore, agave does not require irrigation. However, they do not plan to continue with agave after this first seven-year cycle: *"At first we thought it would be little work but then we realized that it was a lot of work. We knew that a good price is paid for this but now it has fallen. We will change back and sow pasture for cattle on this land, never again will we plant agave."*

Proximate causes: Cattle ranching, market prices, investment capital, irrigation water.

Driving forces: Economic, environmental.

Middle-income household: Rancho de Barra - José Ortiz[30]

José Ortiz, the owner of Rancho de Barra, has extensive migration experience. At the time of the interview in 2004 he was 40 years old. He owns a farm with 4 hectares of rainfed land, 15 hectares of forested semi-arid land in the hills and he also has access to 25 hectares of common land as an *ejidatario*. He was in the United States from 1980-82, 1983-85 and 1987-1993. From 1994 onwards he has spent 3-4 months per year in the US. He went to the US in order to earn enough money for investments he wanted to make in Mexico, especially in cattle and pasture. During his extended migration periods he worked in a chocolate factory and was able to save enough money to open a beauty salon in addition to his factory job. After a few years in the US he was able to save 5'000 to 10'000 USD per year to bring back to Mexico. He used these remittances to purchase land in 1993 and 1999. He also invested remittances in vehicles and agricultural inputs, and bought 40 head of cattle. In order to have enough pasture for his cattle he also used remittances to buy 15 hectares of

---

[29] Name of landowner and ranch have been changed.
[30] Name of landowner and ranch have been changed.

land in the hills, of which he deforested 7 hectares in order to establish pasture for his cattle. He complains about the absence of irrigation water which limits his cultivation choices to rainfed maize, millet or agave. He says a dam should be built to provide irrigation water, which would fundamentally change the situation.

He thinks the number of cattle in his area is lower today than in 1980 due to the economic crisis in the 80s. According to him, today there are more people in the community that are selling cattle than buying additional cattle. His father owned 140 head of cattle from 1980 to 1987. In 1987 he sold most of his cattle and put the capital into the bank, but lost a large amount due to the devaluation of the Mexican peso. He also bought 26 ha of land in 1960, of which 3 hectares are irrigated. In 2001, he decided to change the land use of 16 hectares of maize and sorghum, which were no longer economically worthwhile. He rented this land instead to the company *Azul, Agricultura y Servicios* and kept the remaining area of rainfed maize for his sons' cattle. The terms of his contract with AAS are unique in that he will receive 35% of total income from the harvest but no yearly rent. His other sons bought 25 hectares of sugar cane fields and also own some cattle. Despite concerns about a drop in the price of agave and the difficulties that independent agave producers face in finding a buyer at harvest time, José Ortíz also intended to finance his own agave plantation in 2000 with remittances from the US. However, he realized that it is very expensive to purchase the agave seedlings in order to start a plantation. One agave seedling costs around USD 0.65 (6.5 Mexican pesos in 2003). Around 3000 seedlings are planted on one hectare, which comes to a total cost of USD 1950/ha. Seedlings represent approximately 25% of the total investment to cultivate agave to the point of harvest (Flores and Zamora 2003). He now intends to use cuttings from the agave plantation rented out by his father, and use these for starting his own plantation.[31]

He says that migrating to the United States is a very good opportunity to improve one's livelihood. He describes his stays in the US as "rejuvenating" and says that his experience abroad has given him many ideas on how to do things in agriculture and with regard to irrigation technology and machinery.

Proximate causes: Cattle ranching, market prices, remittances, irrigation water.

Driving forces: Economic, environmental.

---

[31] It is illegal to cut off seedlings on land that is being cultivated by tequila companies. The agave fields are fenced off and "no entry" signs are posted at the gates. Sometimes guards are posted at agave plantations during the night as the stealing of seedlings is a considerable problem for the tequila companies, since they plant valuable improved varieties. During the interview it was not clear whether José Ortíz intended to cut the seedlings despite this formal prohibition.

Analysis of driving forces on private properties

The proximate cause for deforestation on all private properties is cattle ranching, as all three owners require pasture for their cattle. The cattle were financed through capital provided by off-farm activities in two cases and remittances in the other. The fact that capital is invested in cattle is an indication that cattle ranching is good business when pasture is not a limiting factor, which is the case for these private properties.

In all cases, the absence of irrigation water, a proximate cause for environmental factors, makes the change to *agave azul* the economically rational choice. However, if irrigation water were abundant, one landowner would prefer to change to vegetable or sugar cane cultivation instead of agave. This preference is due to the decline of the agave price and the difficulty of finding a buyer as an independent cultivator of agave.

As these are private properties, land tenure is not an issue and was never discussed as a factor influencing land use decision-making.

Overall, the following driving forces and proximate causes underlie land use changes on the private properties:

- **Economic factors**: Cattle ranching, market prices, investment capital, remittances.
- **Environmental factors**: Irrigation water.

7.6 Dynamics of proximate causes and driving forces

Land use changes in the hills as well as on agricultural land are driven by a combination of proximate causes that are all necessary for the change to occur, instead of one single predominant cause. These dynamics can be illustrated with two examples.

Deforestation in the hills is driven by a combination of economic and political-institutional factors. Economic factors play a role, in that cattle ranching and pasture establishment are economically profitable activities. Political/institutional factors also play a role. In one community, the change from common to individual ownership will make it possible for individual landowners to deforest. In the other community it is the absence of regulation together with a certain degree of corruption that drives deforestation. In addition, due to the fact that subsidies can be received for cattle ranching, people have deforested land to create pasture in order to increase the eligible area.

On agricultural land, the absence of irrigation water, rainfall variability and the low market price for maize are the key proximate causes of agricultural change. The two most frequently mentioned reasons why landowners decided to change to *agave azul* were the unprofitability of cultivating rainfed maize and the variability of rainfall and lack of irrigation. Interestingly, the same two factors are mentioned as reasons for maintaining a certain land use system in El Jalocote. The presence of irrigation enabled the cultivation of horticultural crops, which along with sugar cane is a very profitable type of land use. However, there are additional factors driving land use change. The fact that fields planted with agave are also eligible for PROCAMPO subsidies contributes to the decision of landowners to rent out their land. In 1997, average PROCAMPO payments per beneficiary amounted to USD 329 per year (Sadoulet et al. 2001). This represents 46 percent of the gross maize income for a farmer who obtained the average yield of 1 ton per hectare and received an average price of USD 140 per ton. However, without the opportunity of changing to *agave azul*, no land use change would have occurred. This opportunity of converting to *agave azul* is linked to two main driving forces: environmental factors, as decreasing soil fertility and susceptibility to disease have forced companies to search for new land; and globalization processes such as increased market linkages and the steady expansion of tourism which have contributed to boost the international demand for tequila.

These examples indicate that the main driving forces identified in the study area are economic (low market prices for maize, high costs of agricultural inputs), environmental (variability of rainfall, absence of irrigation water) and political/institutional (agricultural subsidies, change in land tenure).

Demography (labor availability) plays an important role in one case study site while technology and culture appear to be less important.

The low and variable rainfall constitutes a central factor in the decision-making of landowners. The overall decrease of rainfall and especially the increased variability was frequently mentioned as one of the main reasons for changing land use from maize to *agave azul*. Cultivating rainfed maize carries a high risk as agricultural inputs and labor are costly and there is a high probability of low and variable rainfall. According to national data on rainfall patterns between 1941 and 2001, the seasonal pattern of monthly rainfall averages in 2003 matches the long-term monthly rainfall averages (**Figure 7-01**). In the region where the study area is located (Region VIII *Santiago Lerma Pacific Region*), average monthly rainfall between March and May was abnormally low between 1990 and 2000 (**Figure 7-02**). While there were frequent positive and negative deviations from average rainfall between 1941 and 1989, rainfall between 1990 and 2000 (with the exception of 1997) was consistently lower than the long-term average.

Figure 7-01 Average, maximum and minimum monthly rainfall in Mexico (1941-2001)

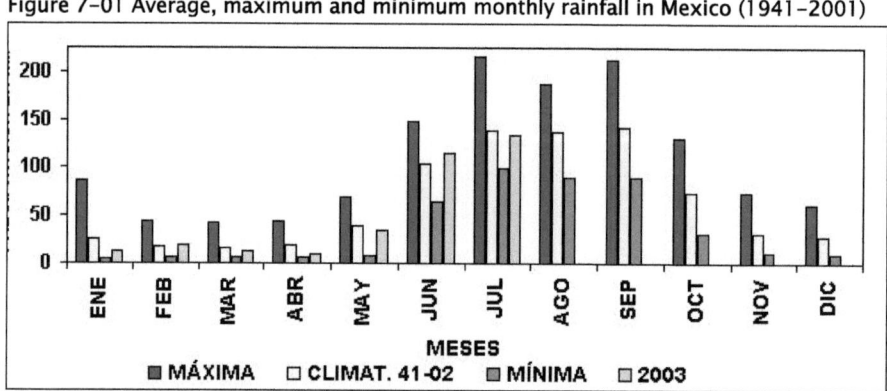

Legend: x-axis=rainfall (mm), y-axis=months
Source: Mexican National Water Commission (Comisión Nacional del Agua. http://smn.cna.gob.mx

Figure 7-02 Rainfall anomaly March–May 1941 to 2004 (Region VIII)

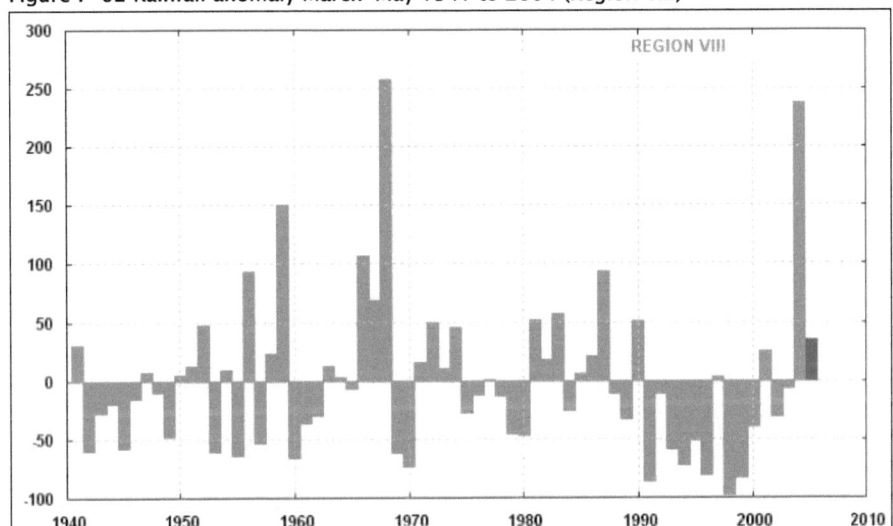

Legend: x-axis = years, y-axis = anomaly of rainfall (in percent). Graph represents rainfall anomaly for the Lerma Santiago Pacific region, which includes the study area. Rainfall anomaly is calculated in percent deviation from average rainfall between March and May each year 1941-2004 (positive or negative). Prognostic for rainfall anomaly for 2005 is indicated in dark grey (9.4mm), 39.1% higher rainfall than average for this time period.
Source: Mexican National Water Commission (Comisión Nacional del Agua).
http://smn.cna.gob.mx/productos/pronostico/elluvias/plluvias.pdf

Cultural factors are thought to play an important role in land use change, but their influence is difficult to identify (Bürgi et al. 2004, Proctor 1998). The complex nature of culture makes it a vague concept, with much disagreement about whether culture encompasses only attitudes, beliefs, norms, and knowledge (Rockwell 1994), or if it includes factors such as population development, economy, technology, and political processes (Proctor 1998). Due to this ambiguity, it is difficult to assign culture an independent direct effect on the environment: *"All scholars working on land use/land cover change grant culture some importance, but most despair of forming any useful generalizations about it, primarily because...those aspects of culture which really matter are the hardest to measure, and culture itself is far too local a phenomenon to be subject to any form of generalization"* (Proctor 1998). Therefore, it is not that culture plays no role, but rather that it is difficult to identify its particular influence. Many proximate causes include cultural components as well as institutional or economic factors. Corruption is one proximate cause that is difficult to attribute to a particular underlying driving force. It is commonly considered a result of policy failure (Geist and Lambin 2002) but can also be due to a difficult economic situation at the national, regional or individual level creating a need for additional income, or can even be considered to be a cultural practice. However, a much more detailed investigation would be required to elucidate the dominant driving forces behind a particular instance of corruption.

Population growth as a proximate cause of demographic factors is not driving land use change in this area as population numbers are actually decreasing due to outmigration. In addition, the absence of labor due to outmigration does not appear to influence land use change in a uniform way. While in two case study sites the role of labor does not seem to influence land use decisions, in Mezquitán a significant number of older people, especially widows, decided to rent out land to tequila companies because they were unable to cultivate the maize fields themselves or to pay for hired labor. The interaction between migration trends and the land use change from maize to agave is therefore ambiguous. Migration may increase the conversion to agave due to the lack of labor for cultivating other crops, and agave cultivation may increase migration because the lower labor requirements free more family members to migrate. However, a number of people think that migration trends decrease in response to the expansion in agave cultivation because tequila companies provide increased opportunities for employment in the region.

The results of the analysis of proximate causes and driving forces in the case study sites are visualized in **Figures 7-03** and **7-04**. **Figure 7-03** gives an overview of proximate causes and underlying drivers for land use change in the hills while **Figure 7-04** gives an overview of proximate causes and underlying drivers for land use change on agricultural land.

Even though this model is based on the assumption that feedback mechanisms exist between land use and proximate causes (see *chapter 4.1*), it was not within the scope of this study to evaluate these feedback mechanisms. Nonetheless, during interviews reference was made to several types of feedback mechanisms, which will be briefly mentioned with regard to the two main types of land use change. There are undoubtedly many more feedback mechanisms that influence land use decision-making, but these will not be discussed here as the data required for a systematic assessment is not available.

Figure 7-03 Proximate causes and driving forces of land use change in the hills

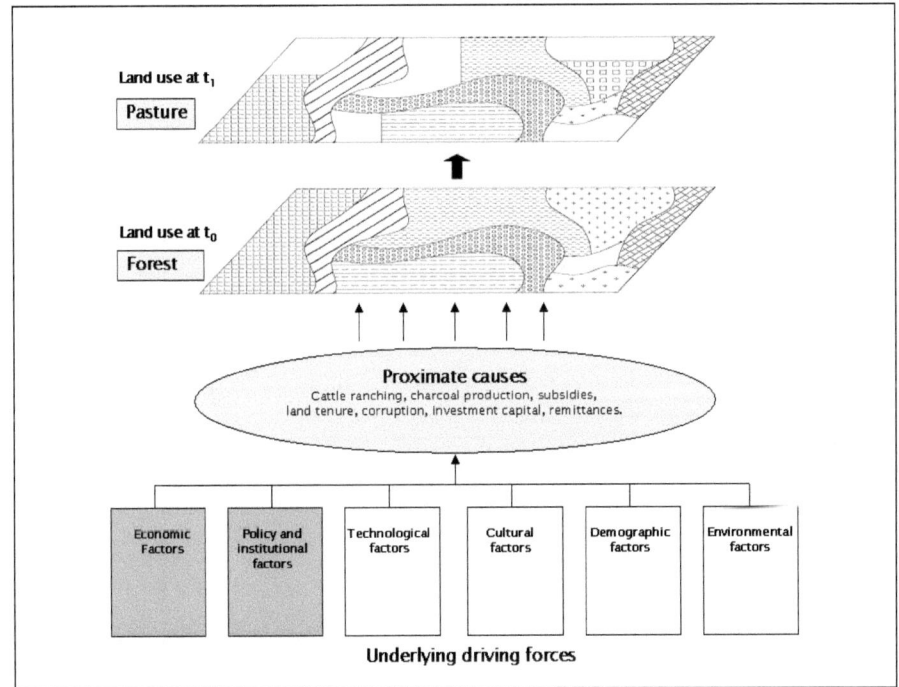

Legend: Proximate causes of the land use change from forest to pasture are marked in light grey while underlying driving forces are marked in dark grey.
Diagram by author.

In 1994, when the PROCAMPO program was initiated, pastures could be registered for subsidies. This motivated landowners to establish new pastures in the early 1990s to gain additional access to subsidies, which contributed to deforestation in many parts of Mexico (Turner et al. 2001). Klepeis and Vance (2003) developed a model that suggests that each additional hectare subsidized reduces forested area by 0.41 ha. The authors concluded that the unintended effect of PROCAMPO on increased deforestation is substantial. Deforestation is believed by many people in the region to cause changes in the local climate, in particular lower and more variable rainfall, which in turn negatively affects agricultural production on rainfed land and thus increases dependence on other sources of income such as livestock and the concurrent expansion of pasture to support them.

Figure 7-04 is based on the land use modification of maize to *agave azul* because it is the main land use change occuring in the region. Other land use changes are mostly limited to changes between horticultural crops.

117

Figure 7-04 Proximate causes and driving forces of land use change on agricultural land

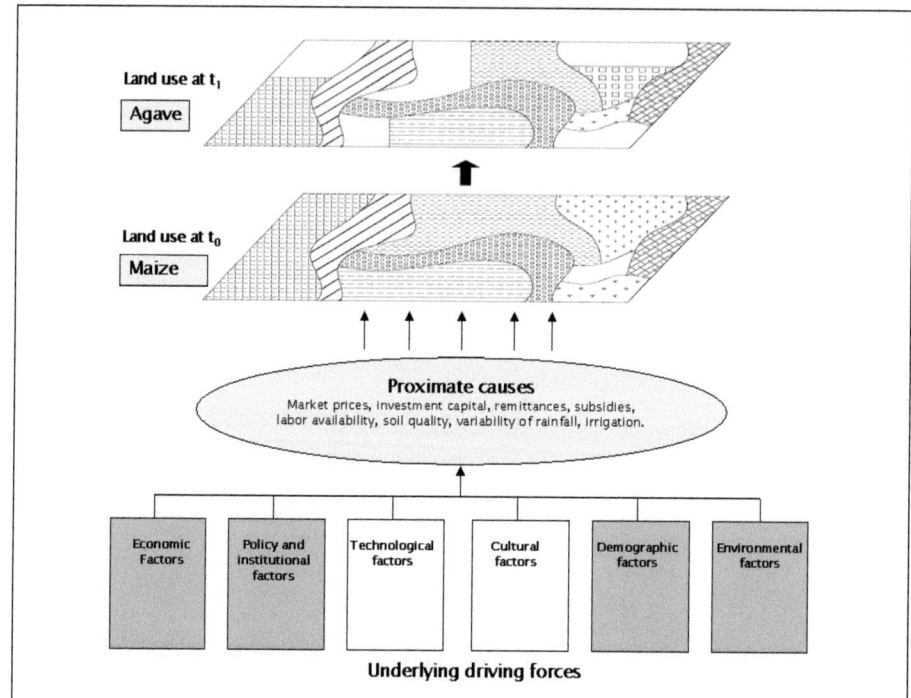

Legend: Proximate causes of the land use change from maize to *agave azul* are marked in light grey while underlying driving forces are marked in dark grey.
Diagram by author.

The rapid expansion of *agave azul* has led to a steady decrease in market price. This trend could slow the land use conversion of maize to agave or even incite some landowners to switch back to maize cultivation. However, the high application rates of pesticides and herbicides required for agave cultivation are leading to a higher level of chemical soil degradation than under traditional maize cultivation. This feeds back into land use decision-making in at least two ways: tequila companies renting land might leave the region when harvests decrease due to depleted soils, and the landowners might not have another option than to continue with agave cultivation as it may not be possible to cultivate maize on the depleted soils.

7.7 Conclusions

The main differences between land use changes occuring in the hills and on agricultural land relate to land tenure and the importance of environmental factors for agricultural land. The case examples demonstrated that common or unregulated land tenure can influence land use on communal land in the hills, which is not the case for agricultural land belonging to individual landowners. On the other hand, environmental factors such as rainfall, availability of irrigation water, topography and soil quality are much more important for agricultural land than for the land in the hills which is generally used for pasture or shifting cultivation. The planting of pasture is inexpensive to initiate and maintain, both in terms of labor and chemical inputs. PROCAMPO stipulates that eligible land must be maintained under production, and fallow land does not qualify for support. The planting of pasture thus offers a low-cost option of ensuring continued support from PROCAMPO with little concern for the actual productive value of the land (Klepeis and Vance 2003). This is very different than agricultural land where inputs are costly and the labor requirements considerable. To recover investments, the influence of environmental factors on the productivity of agricultural land is decisive.

Overall, the main driving forces identified in the study area are economic, political/institutional, demographic and environmental. Land use change is not driven by one single factor but by a combination of factors that are all necessary for the change to occur. Technology is less important, and the influence of culture is notoriously difficult to evaluate. Remittances were found to impact land use change in all case study sites, but only to a certain extent. The impact of remittances depends to a large degree on the socio-economic, political and environmental context of the community and on the individual situation of the household. This will be discussed in the next chapter.

8. The impact of transnational migration: Remittances as drivers of land use change

8.1 Characteristics of transnational migration from Mexico to the US

Migration is increasing continually in Mexico (Goldring 2004, Verduzco and Unger 1998). During 1971-1980, 640'294 legal Mexican migrants arrived in the US, a figure that more than doubled over the next decade (1981-1990), reaching 1'655'843. These numbers do not include unauthorized Mexicans living in the US, which are difficult to measure but were estimated at 4 million in total in 1980, increasing annually by approximately 200'000[1] (Bean et al. 1998). The total Mexican population in the US is estimated to reach 25 million in 2010 (Bean et al. 1998). With the steady flow of outmigration, population is actually declining in all of the research case study sites and this trend is expected to continue (Municipality of Autlán 2003). As it is mainly young people who leave, the overall population is aging and the fact that mainly old people, women and small children are left behind in the communities is often a subject of discussion and concern.

The large majority of migrants are young men between 18 and 30 years old who all migrate to the US, mainly to California but also to Minnesota, Oregon, North Carolina and several other states. Women rarely migrate; if they do, it is usually to join their husbands abroad. While some migrants in the study area benefited from the general amnesty legalizing their migrant status in 1986[2], approximately 90% of the migrants included in this survey were undocumented. Since controls at the US-Mexican border have intensified over the years, the illegal crossing of the border is becoming even more dangerous and many people have died in the attempt to walk across the desert or cross rivers. As a consequence, periods of migration are becoming longer due to the danger and expense of frequent border crossings. Hence, men often do not come home for many years at a time.

Local opinions about the benefits of migration are divided. Many consider migration as positive because remittances constitute a crucial source of income for many families. However, migration is also considered to be a negative phenomenon – people cannot make a living in their native country, so they have to go to a foreign country to find work. The fact that so many men have to leave their families behind in order to secure a livelihood gives an indication of the difficult situation in the communities of origin.

---

[1] Estimation is based on the years 1986-1990 (Bean et al. 1998).
[2] In 1986, the United States Immigration Reform and Control Act (IRCA) offered legal status to approximately 2 million undocumented Mexican migrants who had been either working in agriculture or living illegally in the United States since before 1982 (Bean et al. 1998).

On the left: Migrant husband who has returned to his family after a short migration period of 7 months in 1993. On the right: Wife and daughter of migrant husband who has been working in the US since 1994. Since 2000, he has been able to visit his family in Mexico once a year. Photos by author (March 2004).

A substantial amount of money is necessary in order to migrate. Currently, between 2'500 and 3'500 USD3 is paid to people (known as coyotes) who bring migrants illegally across the border. The necessary funds are often borrowed from relatives or friends who have already migrated to the US. Another important factor for successful migration is access to a network of people in the US, to have a first "stepping stone" upon arrival (Massey and España 1987). This network often helps newly arrived migrants locate housing and find a job. Migrants find work primarily in the agricultural sector, on construction sites, in factories or in restaurants (see **Table 8-01**). The occupation in the destination country reflects the nature of the network the migrants belong to. In El Jalocote, the migrant networks are mostly organized by family. If one family member has managed to establish himself with a place to live and a steady job in the US, he acts as a stepping stone for other members of his family. For example, if the first migrant of a family works in construction, then he can often secure a job for another arriving migrant family member who in turn then "sponsors" the next migrant, such as a younger brother. This leads to situations like that of one family from El Jalocote, where all the migrated family members work in the same restaurant in Oakland, or another family where the migrant members work in factories in Minnesota producing

---

[3] The amount paid depends on the starting and destination point of the journey and whether a job is secured in advance at the destination (Interview data, 2004).

ship parts. In contrast, in Chiquihuitlán the majority of migrants engage in seasonal migration to pick fruit for six months of the year in California. This pattern of migration from Chiquihuitlán to California started more than 50 years ago. Finally, most migrants from Mezquitán find work in the US on construction sites and in factories.

Table 8-01 Occupation of migrants in the US (%)

Occupation of migrants in US	El Jalocote (24 migrants)	Chiquihuitlán (11 migrants)	Mezquitán (45 migrants)	n = 80 Total (%)
Paid agricultural labor	29	73	4	21
Construction sites, factories	17	18	67	45
Wife joining husband	25	0	9	13
Restaurants	17	0	11	11
Househelp, maid, gardener, childcare	12	9	9	10
Total	100	100	100	100

Source: Interviews by author.

The main reason of migration is the absence of job opportunities at home (47%) and the desire to earn more money that would enable the family to make special investments (27%) (see **Table 8-02**). Even if the family owns land, the income to be earned through agriculture is often not enough for a secure livelihood. As off-farm jobs are rare, migration is often the only option for earning enough money to even cover basic subsistence needs. Furthermore, a very important motivation is the dream of building a house. For many families, it is only through migration that they are able to accumulate enough capital to purchase the building materials for a new house or to repair an existing one. A few men migrate in order to save enough money to be able to open a small business upon return. Nonetheless, most people know very well that the cost of living in the US is also much higher and that it is difficult to save significant amounts of money. Women migrate almost exclusively in order to join a husband working in the US. Lastly, some young men are attracted to migration because of the sense of adventure that is involved in going to "El Norte."[4]

---

[4] "El Norte" [The North] is a popular designation for the United States.

Table 8-02 **Reasons for migration (in %)**

n = 60	El Jalocote	Chiquihuitlán	Mezquitán	Total
Couldn't find work here [in Mexico]	57	29	48	47
More money (to be used for special investments) can only be made in the US	17	57	17	27
To join other family members already in migration in the US	17	0	0	7
Prefers to live and work abroad	9	14	26	16
Other	0	0	9	3
Total	100	100	100	100

Source: Interviews by author

Approximately half of all households in the case study sites have had members in migration in the US at some point since 1980 and are therefore designated as "migrant households". In El Jalocote 54% are migrant households, in Chiquihuitlán 55% and in Mezquitán 41%. On average, migrant households have 1-2 family members in migration. Only 11 to 22% of migrants return permanently to Mexico after seasonal or temporary migration periods. As can be seen in **Table 8-03**, the situation in Chiquihuitlán is different from the other two communities because on average a migrant household has only one family member in migration and 66% of migrants engage in seasonal migration, staying six months per year. This reduces the average duration of migration period to 1.5 years, much shorter than migration periods in the other two case study sites (7.7 and 8.4 years). During the interviews, it was frequently mentioned that the family member left with the idea of saving enough money to come back and construct a house. In reality, very few migrants return to the community to live. The typical case is that the migrant finds steady work and then either marries someone from the US or his wife/girlfriend joins him from Mexico. Once they start a family they are very unlikely to return to Mexico, and often reduce the amount they send to relatives or stop sending money altogether.

Table 8-03 **Migration characteristics**

	El Jalocote	Chiquihuitlán	Mezquitán
Average number of migrants per family	2	1	1.7
Average duration of migration period in the US	8.4 years	1.5 years[5]	7.7 years
Migrants definitely returned to Mexico (in % of total migrants per case study area)	22%	25%	11%

Source: Interviews by author.

---

[5] This average is markedly lower because 66% of all migrants engage in seasonal migration and stay six months per year.

Reasons for not migrating are similar for all case study areas (**Table 8-04**). The most common reason for not migrating is that all family members have work, either on their own farm or in government or local industry. In Chiquihuitlán the main reason for not migrating is a lack of capital for the illegal border crossing (43%). This again reflects the higher level of poverty in Chiquihuitlán compared to the other two communities, where only 13% and 3% of households indicated the lack of capital as the reason for not migrating. The other main reasons cited for not migrating are the danger involved in illegally crossing the border; a father who does not want to leave his family alone, especially if he has small children; and the simple preference to live and work in Mexico.

Table 8-04 Reasons for not migrating (%)

Reasons for not migrating	El Jalocote	Chiquihuitlán	Mezquitán	Total
Family has work here [in Mexico]	25	29	36	33
The household does not have the money that would allow somebody to leave	13	43	3	9
Household members prefer to live and work in Mexico	13	0	18	15
The danger of crossing the border (illegally)	13	14	15	15
The daily life in the US is very difficult	0	0	5	4
The family will not allow the person to leave	0	0	3	2
Family father does not want to leave family alone	25	14	13	15
Other	11	0	7	7
Total (n=41)	100	100	100	100

Source: Interviews by author

In this study, contrary to expectations, the loss of labor force due to migration of family members was not always perceived negatively by households. Overall, the working load for women appears to change little due to migration of family members. Working load of men and women worsened in all the case study sites in similar (limited) proportions. Interestingly, while in Mezquitán the working load improved for 8% of the women, but for none of the women in El Jalocote and Chiquihuitlán, the situation is different for men. Migration of household members actually improved the working load for 8% of the men in Mezquitán, for 11% of the men in Chiquihuitlán and for 27% of the men in El Jalocote. The main reason cited for this improvement was that in these families remittances sent by the children improved the economic situation of the household to such an extent that the men of the household were not required to work as hard as before. The lack of improvement in women's working load might be explained by the fact that women are usually responsible for household tasks and the caring for children. Even though the household income might increase with remittances, they continue to work as hard or even harder with members of the family in migration.

Table 8-05 Changes in work load for members of migrant household (in %)

Working load for women	El Jalocote	Chiquihuitlán	Mezquitán
Better	0	0	8
No change	64	89	69
Worse	36	11	23
Total	100	100	100
Working load for men			
Better	27	11	8
No change	46	78	69
Worse	27	11	23
Total (n=46)	100	100	100

Source: Interviews by author.

In most cases (see **Table 8-08**), households with migrants report to be in a much better financial situation than before a member of the family migrated. Wealthy families constitute an exception as they do not need further income even though they have the necessary capital to migrate. Members of wealthy families travel to the US for vacation or to visit family members, or send their children for higher education. For poorer and middle class families, migration diversifies and augments their existing sources of income, which is often crucial for meeting subsistence needs. The consequences of migration are highly variable. While some migrants become wealthier, others only just survive. A number of families suffer from a "failed" migration attempt due to large loans that were taken in order to finance migration. These loans heavily indebted the family and will be difficult to pay back, especially without remittances. The situation of these households actually worsened after attempted migration, sometimes dramatically so. Nevertheless, the risk of heavy indebtment is taken in the hope of obtaining a well-paid secure job in the long term, which is a very unlikely prospect in the home country. In all case study sites, the very poor families cannot migrate as they lack the minimum level of material assets necessary for migration.

Many women raise their children alone while their husbands migrate. Most husbands only return every 5-8 years and some never return because they have started another family in the US. In communities with high migration activity, young men are largely absent. During the interviews many women mentioned that the absence of the husband/father was difficult for them and their children because they have to take on the family responsibilities and household tasks that would normally be done by the husband.

Every person interviewed thinks that migration rates will increase in the future. It continues to be difficult to make a living by staying in the communities, and many young people do not want to work in agriculture. Since local employment opportunities are limited, migration becomes their main livelihood strategy, and they leave with the aim of never coming back.

8.2 Impacts of transnational migration on land use changes

There are a number of ways that migration influences land use and land use influences migration. One of these interactions is the influence of land use change from maize to *agave azul* on migration. Changing land use from maize to agave allows families to rent out their land to tequila companies and frees family members to migrate to the US, thereby diversifying and increasing their sources of income. The arrival of tequila companies renting land in the region is encouraging migration because in reverse leasing arrangements, the land is rented for cycles of seven years, paying yearly rent to the owner and 3 to 5% of the harvest. This allows land owners to receive a yearly income from land rent and at the same time migrate to work in the US. Anecdotal evidence indicates that one of the reasons so many migrants come home at Christmas is because that is when the tequila companies pay out the annual rent. Land that is cultivated with agave is also eligible to receive PROCAMPO subsidies. The subsidy is paid to the landowner even if the land is rented out to tequila companies. This means that by renting out land for agave the household has a triple source of income: 1) Yearly rent for the land (between 100-130 USD/ha/year,[6] 2) PROCAMPO subsidies (approximately 120 USD/ha/year), and 3) Income from other employment, often as an agricultural worker for the tequila company renting the land, or from migration.

Vice versa, migration also drives the land use change from maize to agave. In the *ejidos* many land owners are elderly. *Ejidal* land is often passed on to a son or a daughter late in the life of the *ejidatarios*.[7] For instance, in the *ejido* La Canoa, also located in the municipality of Autlán, 84% of *ejidatarios* were 51 years or older in 1993 (Nujiten 2001). As the *ejidatarios* get too old or sick to farm their land, and with children often in migration abroad who cannot help them with farming, renting out their land to the tequila companies is often a welcome opportunity. They will have two sources of income: PROCAMPO and the yearly rent, which is economically more profitable than hiring labor and investing in the necessary agricultural inputs to cultivate rainfed maize. Additionally, remittances sent by their children often constitute a third source of income.

---

[6] The rent paid depends on the location of the land plot, on the size (the larger the total land plot, the higher the rent per hectare) and on the quality of the soil (Vargas Martin 2003).
[7] The agrarian law states that the *ejido* plot cannot be divided and that the agrarian right has to be left to one heir. The *ejidatarios* can designate the heir to the agrarian right from amongst their partner and children (Nujiten 2001).

One of the questions examined in this study was whether people with land are more likely to migrate than landless households. Land ownership was not correlated with having household members in migration (**Table 8-06**). The relative economic situation of the individual household has a stronger influence on whether members of the family will migrate than does whether or not they own land. Several factors lead members of the household to migrate: 1) In order to enable members of the household to migrate, the household must either have the financial capital for migration or relatives or friends who will lend them the money; 2) The person wishing to migrate has not found satisfactory employment locally; and 3) There must be a need for additional income, pushing members of the family into migration. The importance of these factors holds true for non-migrants, because either they do not have sufficient capital to migrate, they have enough work in Mexico, or they do not need to earn more money (see **Table 8-04**). There are few cases where young men migrate simply because they prefer to live and work abroad or for the sense of adventure involved.

Table 8-06 Differences in migration activity between landless and landowning households (in %)

Ejido	Percentage of migrant households of total landless households (n=44)	Percentage of migrant households of total landowning households (n=43)
El Jalocote	33	75
Chiquihuitlán	100	40
Mezquitán	40	60

Source: Interviews by author.

8.3 Investment of remittances

In the case study sites, 87% of migrant households receive remittances. This income is mainly sent by money order. Some families have bankcards that allow them to withdraw money from an ATM in Mexico from the bank account that a family member has opened in the US. Survey results showed that the amounts remitted vary from 60 USD to 6000 USD per year, averaging 1562 USD per year per migrant. However, remittances are difficult to estimate due to the irregularity with which they are sent, and often no money is remitted during the first few months (or even years). The importance of remittances as compared to other sources of income at the household level was assessed using the percentage that remittances represent of total income as an indicator. **Table 8-07** shows that on average for 20% of migrant households, remittances represent the largest source of income, and for an additional 15%, the second most important income after agriculture. As can be seen in this table, remittances are very important as they represent approximately 50% of total household income in El Jalocote and Chiquihuitlán.

Remittances are less important in Mezquitán where remittances represent on average only 28% of household income, confirming again the relatively prosperous situation of this community. In the poorer communities remittances may constitute up to 90% of a household's income.

Table 8-07 Significance of remittances as source of income (in %)

(n=46)	El Jalocote	Chiquihuitlán	Mezquitán	Average
Remittances as a percentage of total household income (average of all migrant households)	48	59	28	45
Percentage of households for which remittances is largest source of income	11	33	15	20
Percentage of households for which remittances is second largest source of income	7	22	15	15

Source: Interviews by author.

Table 8-08 gives an overview of the impact of remittances on the economic situation of the household, the living conditions, and the education of family members. The economic situation has improved in all *ejidos* due to remittances, with the highest impact felt in Chiquihuitlán, where 89% of all migrant households report an improvement of their economic situation. In El Jalocote and Mezquitán the impact is not perceived as strongly, but still 64% and 54% of households report an improvement. Living conditions, such as the state of the house, furniture and electrical appliances have also improved for more than half of the households in El Jalocote and Chiquihuitlán. This is the case for only 15% of the households in Mezquitán. This is surprising because in Mezquitán many migrant households are investing remittances in house construction and could therefore be expected to report better living conditions. However, this was not the case. Perhaps since their existing housing is already of quite good quality, the difference to even better housing is not as marked.

The impact of remittances on education is quite low. Only 11 to 27% of households report an improvement of the education level of family members due to remittances. This is probably due to the fact that good quality or higher-level schools are only to be found in the town of Autlán, which would require a considerable investment to cover school clothes, food, and travel expenses. As on average only 10% of remittances are invested in the education of children, the impact is rather limited.

Table 8-08 Impact of remittances on situation of household (in %)

n=34	El Jalocote	Chiquihuitlán	Mezquitán
Economic situation of family			
Better	64	89	54
No change	36	11	46
Worse	0	0	0
Total	100	100	100
Living conditions (Quality of housing, furniture, household appliances)			
Better	55	56	15
No change	45	44	85
Worse	0	0	0
Total	100	100	100
Education of family members			
Better	27	11	15
No change	73	89	85
Worse	0	0	0
Total	100	100	100

Source: Interviews by author.

A wealth ranking of the three wealthiest families of each community was done independently by three different persons. The ranking was identical with the exception of one rating in which the third wealthiest family of one community was different. It is interesting to note that besides having a good sense for business and working hard, the same three factors were mentioned as to why these families are wealthy: they own land, they own cattle and they have family members that are in migration in the United States.

Even though there are several Mexican migrant associations in the US channeling communal remittances to their home communities in the state of Jalisco, there are few communal remittances to the case study sites.8 All people interviewed agree that remittances mostly benefit the individual family and not the community as a whole. There have been few cases where remittances have been invested to improve the situation of the community, for example by creating jobs or by improving infrastructure so that the entire community would benefit. Nonetheless, in El Jalocote it was

---

[8] Research in the community of Rincón de Luisa showed that migrants from this community had formed an association in Las Vegas and regularly send money for community projects such as the renovation of the church, constructing a new "plaza" (main square), a community garden, etc. (Portner 2005).

mentioned that the presence of a shop was very useful and that the establishment of this shop was only possible because the owners returned from a long migration period in the US with the necessary capital.

As can be seen in **Figure 8-01**, in the case study sites the majority of remittances are used to cover subsistence needs (food 21%, medical costs 15%), to pay back loans (15%) and for house construction (14%). A smaller part of remittances are invested in the education of children (9%), electronic appliances (8%), savings (6%) and agricultural inputs such as fertilizer and pesticides (5%). An additional 2% of remittances are invested in land purchase and agricultural tools. If investment patterns are examined from the angle of productive and non-productive investments, it can be seen that 15% of remittances are invested in productive activities such as agricultural production, land purchase, vehicles and business investments. This represents a similar level of productive investment compared to studies conducted in a number of developing countries where productive investement is situated around 20% (Sander 2003). However, results depend strongly on the method used to distinguish between consumer and investment spending. 9

---

[9] The difference between productive and non-productive investments is not easily made. This is discussed in more detail in *chapter 3.3*. In this study, investment in agricultural tools and inputs, land purchase, vehicles, business investments and savings are considered productive investments while investment in food, health, education, paying back loans, house construction, migration costs of family members, electronic appliances, telephone cards and travel are considered unproductive investments.

Figure 8-01 Investment of remittances (in %)

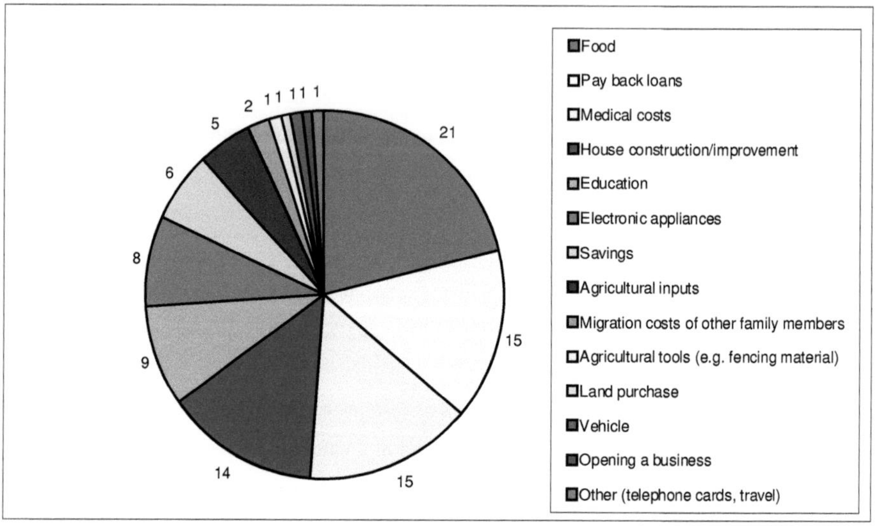

Segments (with legend): Food 21; Pay back loans 15; Medical costs 15; House construction/improvement 14; Education 9; Electronic appliances 8; Savings 6; Agricultural inputs 5; Migration costs of other family members 2; Agricultural tools (e.g. fencing material) 1; Land purchase 1; Vehicle 1; Opening a business 1; Other (telephone cards, travel) 1.

Source: Interviews by author.

8.4 Impact of remittances on land use change

In order to evaluate the impact of remittances on land use change, a number of indicators were selected according to the concept of remittance landscape presented in *chapter 4.2* Remittances are found to directly influence land use change if one or more of the investments listed in **Table 8-09** are made. There are five different types of investments leading to land use changes. **Table 8-09** indicates the number of households (in %) of all households that receive remittances and who make one or more of these types of investments.

Table 8-09 Impact of remittances on land use changes (in %)

n = 42	El Jalocote	Chiquihuitlán	Mezquitán
1. Remittances are used to buy material to fence off land in the hills to use as pasture, either for own cattle or to rent out.	9	22	0
2. Remittances are used to buy cattle for which new pastures are opened in the hills.	9	22	0
3. Remittances are used to buy land and land use of this land is changed.	9	0	9
4. Remittances are invested to start an agave plantation, changing land use from maize to *agave azul*.	0	0	9
5. Remittances are invested in house construction.	27	22	32
Number of households of all remittances-receiving households (in %) investing remittances in categories 1 to 5 leading to land use changes.	27	22	41

Source: Interviews by author.

22% of all remittance-receiving households in Chiquihuitlán and 9% of all remittance-receiving households in El Jalocote have invested remittances in cattle, while none have done so in Mezquitán. The same investment pattern is observed for fencing material used for fencing off common land for subsequent use as pasture. In El Jalocote and Mezquitán, 9% of households have used remittances to purchase land while none have done so in Chiquihuitlán. Only migrant households in Mezquitán have used remittances to start their own agave plantations. Many households receiving remittances invest in improving their house, either repairing or enlarging it, which is not considered an investment that leads to land use changes. In contrast, between 22 and 32% of all households use remittances for constructing a new house.

In Mezquitán a visible construction boom is taking place. The construction of new houses involves a land cover change from agricultural to urban land as almost all of these land plots were previously cultivated. Furthermore, the increased density of the villages and the emergence of new, large and sometimes even two-story houses leads not only to a land use change but also to a visual impact at the landscape level. In El Jalocote and Chiquihuitlán, construction activity is also visible, but very few families are building exceptionally large and ostentatious houses.

An example of the many houses that are being constructed in Mezquitán. Photo by author (April 2004).

The investment of remittances in ways that affect land use change is not the dominant use of remittance income, as in all study sites less than 50% of households invest remittances in these ways. The primary type of investment is in house construction (22 to 32% of all households). However, in Mezquitán overall investment of remittances leading to land use changes concerns 41% of all households. Furthermore, even though only 9 to 22% of households invest in fencing material and/or buy cattle, each area affected by the land use change from forest to pasture is large, ranging from 10 to 40 ha. The percentages are also rather low due to the fact that all remittance-receiving households were included, even those who do not own land and are therefore less likely to buy cattle or start cultivating agave. Very few households use remittances to purchase land. This is to be expected due to the land tenure system, as selling and buying *ejidal* land - although it did occur - was illegal before 1992.

The indirect influence of remittances also needs to be considered. Several families stated during interviews that remittances allowed them to cover household expenses, freeing other sources of income for investment that **does** influence land use change. For instance, remittances may be used to cover daily living expenses while other income is used to buy goats and graze them on common land, or to establish agave plantations. According to the interviewed families, even though the land use change is not caused by the direct investment of remittances, it would not have occurred without remittances. Many of the land owners in Mezquitán who started to cultivate agave independently were able to do so because of a long-term, secure off-farm job providing them with regular income. However, remittances are also an important driver. Bowen (2004) found that 20% of the independent agave producers financed a majority of their agave production using remittances and an additional 30% depended on remittances to cover other expenses, freeing capital to be invested in agave cultivation.

Besides the common investment of remittances in housing, there are clear differences in the way remittances affect land use in the three communities. These differences are due to the different socio-economic and environmental contexts of the communities (**Table 8-10**). The community of Chiquihuitlán owns much more communal land (11'598 ha) than El Jalocote and Mezquitán (962 ha and 562 ha respectively) (SAGARPA 2006). This suggests that investment in cattle makes sense due to the availability of pasture, which is often the limiting factor for engaging in cattle ranching in other areas of the municipality. In addition, remittances are invested in fencing material to fence off parts of the common land for use or renting out as pasture. Conflicts around land tenure issues are said to be numerous in Chiquihuitlán (Blanco Barbosa 2004, Martinez 2004, Medina 2004), and a powerful group of *caciques* control access to common land, which is often linked to corruption. Even though there is no formal evidence, it seems likely that the fact that the land regularization program PROCEDE has not been conducted in this community along with endemic corruption and *caciquismo*, provide an encouraging environment for illicit investments in fencing pasture. *Caciques* are often affiliated with the PRI (*Partido Revolucionario Institucional*; Institutional Revolutionary Party), which is the largest conservative political party in Mexico. This political alliance has influenced the allocation of communal land. The *caciques* and their relatives have used their control over the directive board of the community for their own benefit. They have facilitated access of allies to communal land while opposing access to families belonging to the democratic party PRD (*Partido Revolucionario Democrático*; Democratic Revolutionary Party) (Gerritsen 2002). During the interviews, it was mentioned several times that *caciques* pay the elected official in the community to give them land, and that nothing is ever done about it. In El Jalocote, the same type of investments in fencing material were made, but within the legal framework. The families who fenced off land in the hills either had the right to use the communal land or were the legal owners of the land, with the corresponding land titles as established by the PROCEDE program. In Mezquitán, this kind of investment in unauthorized fencing of communal land does not take place because access to the common land is much better regulated. Furthermore, individual land distribution of common land is imminent, so illegally invading common land shortly before individual land titles are assigned makes little sense.

In order to address the question of to what extent the socio-economic,[10] political and environmental context influences the investment of remittances, **Table 8-10** gives an overview of the main differences in context and investment patterns of remittances. Some *ejidatarios* in Mezquitán own irrigated sugar cane fields in the communities of El Volantín and Chacalito which are very profitable, more so than *agave azul*. This favorable natural environment together with the fact that the government-run *Comisión Federal de Electricidad* employs a significant number of people in Mezquitán is to a large extent responsible for the relative wealth of Mezquitán. Infrastructure, housing, and its location on a main road facilitating access to markets, schools and hospitals is much better than in the two other *ejidos*. Even though many *ejidatarios* in El Jalocote also own irrigated land, their land plots are much smaller than those of the *ejidatarios* in Mezquitán. Furthermore, the isolated location of El Jalocote makes transport and access to markets much more difficult.

According to the New Economics of Labour Migration theory (Taylor 1999) (*chapter 3.4*) it would be expected that the favorable context of Mezquitán would mean more investment opportunities and lead to a higher investment of remittances in productive activities compared to the investment patterns of the other two communities with less favorable contexts. However, the only evidence supporting this theory is the fact that in Mezquitán 9% of households have invested in becoming independent agave producers. Even though 9% of households in Mezquitán have also bought land with remittances, this is also the case for El Jalocote. This result is in contradiction to NELM theory, as the socio-economic context in El Jalocote is much less favorable than in Mezquitán.

---

[10] In order to assess the socio-economic context of the community (the level of poverty), three indicators were used: 1) Number of people in the survey indicating PROGRESA (a direct payment program targeting low income households) as first or second largest source of income in the survey; 2) State of housing in the community (% of brick housing, number of houses under construction, state of existing housing); and 3) Information from key informants.

Table 8-10 Main differences in context and investment patterns of remittances between case study sites

	El Jalocote	Chiquihuitlán	Mezquitán
Environmental context	Irrigated agriculture in hilly terrain. Almost all landowners own an irrigated land plot. El Jalocote provides part of drinking water to the town of Autlán.	No irrigated agriculture. Agricultural rainfed land is located on mainly flat land in the valley.	Partially irrigated agriculture. Agricultural land is located on flat to slightly sloping land.
Infrastructure	School, health clinic, two small shops, one cobblestone street. Mostly brick houses, some adobe houses.	School, health clinic, one small shop, one short cobbled street. Mostly adobe houses.	School, health clinic, several small shops, one large restaurant, most streets are made of cobblestone. Mostly brick houses. Many new houses are being constructed.
Location of community	In the hills, at the end of a valley on a dead-end dirt road. Approx. 40 minutes from next town and market.	At the end of a valley on a dead-end dirt road. Approx. 20 minutes from next town and market.	Located on the main road linking Guadalajara to the Pacific Coast. Approx. 20 min. from next town and market.
Socio-economic context	Rather poor community. Mainly smallholders and day laborers in local horticulture industry. Some women work as domestic employees in Autlán.	Poor community. Mainly smallholders working as day laborers in horticulture or in construction sector.	Prosperous community. Many families have diversified income portfolio because family members are employed by the federal electricity company or have other off-farm employment besides farming or renting out own land.
Political context	Participated in the PROCEDE program in 2002. Individual land titles distributed.	Will not participate in PROCEDE. Many conflicts linked to land tenure.	Participated in the PROCEDE program in 1999. No apparent land tenure conflicts.
Average size of: – land plots – *cerro* land	1.5 ha irrigated 7 ha	2.4 ha rainfed 16 ha	8.3 ha irrigated & rainfed 9 ha[11]
Remittances as average percentage of total income	48%	59%	28%
Average yearly money sent to migration household (USD)	780	2095	1809
Investment of remittances (in order of importance)	1. Food 2. Debt repayment 3. Medical costs 4. Agricultural inputs and tools 5. House construction and improvement	1. Food 2. Debt repayment / medical costs 3. House construction and improvement 4. Electrical appliances 5. Agricultural inputs and tools / education	1. Food 2. House construction and improvement 3. Medical costs 4. Debt repayment 5. Education

Source: Interviews by author.

---

[11] As mentioned in *chapter 6.3*, common land is not yet distributed in Mezquitán. Each *ejidatario* and each *ejidataria* will receive approximately 9 ha.

8.5 Conclusions

It appears that migration success is largely influenced by the following two key issues: the capital resources required to reach the destination country, and access to a social network in order to get established and secure a job. Furthermore, the two factors are linked. Those who are part of a social network have relatives or friends within this network who often also lend them the capital for initial migration. Once the new migrant has gotten established in the destination country and paid back the loan, he can then finance the migration of another member of the family and so forth. For those households who do not have access to either capital for migration or to social networks, migration is not a viable option even though it might be desired.

The land use change from rainfed maize to *agave azul* illustrates how migration can influence and is influenced by land use changes. The system of cultivating *agave azul* in reverse leasing arrangements frees labor resources and may therefore accentuate migration movements. On the other hand, limited labor availability due to family members in migration may accelerate a land use change if that change has lower labor demands. This is the case for *agave azul* compared to the cultivation of maize or sorghum on rainfed land.

The impact of remittances depends on the socio-economic, political and environmental context of the community. In Chiquihuitlán, where the environmental (sufficient pasture available) and political (opening of pastures on common land tolerated) context is favorable for cattle ranching, 22% of households have invested in buying cattle whereas only 9% have done so in El Jalocote and none in Mezquitán where the context for cattle ranching is not as conducive.

Differences in the investment patterns of remittances do not provide strong support for the theory of New Economics of Labour Migration (NELM), which stipulates that a higher proportion of remittances are invested in productive activities in areas with favorable socio-economic, political and environmental context. While it is true that only migrant families from the relatively favorable context of Mezquitán invested in more profitable land use systems, investment in land purchase, agricultural tools and business investments were not higher than in the other two communities.

Remittances were found to lead to the following land use changes: 1) forest to pasture, 2) agriculture to urban, and 3) change of agricultural system (subsistence crop to cash crop). The change of agricultural systems is not included in the original concept of remittances landscape as presented in *chapter 4.3* because the basic landscape type remains unchanged. In view of the important socio-economic and environmental changes that accompany the land use change from maize to *agave azul*, it should be considered as a potential type of emerging remittance landscape, if driven mainly by the investment of remittances. Yet, in these case studies the extent to which remittances drive the expansion of *agave azul* is limited. Only in Mezquitán, where 50% of independent agave producers use remittances directly or indirectly for agave cultivation, the emerging landscape can be classified as a remittances landscape. In the other areas even though the investment of remittances is reflected in changes in the landscape, a large majority of agave is planted by commercial agave companies.

PART III

9. Synthesis

9.1 Driving forces at the global level

Globalization involves increasing international capital flows, growing migration and expanding information and communication (Bolay 2004). International trade agreements such as the North American Free Trade Agreement (NAFTA) are a central component of globalization, increasing the international integration of production and consumption markets, often in previously marginal countries. NAFTA, which was concluded between Canada, USA and Mexico in 1994, has been a significant driving force of land use change in the study area by influencing market prices of maize. As a result of NAFTA, the real price of maize in Mexico dropped by 46% between 1994 and 2004 (Eakin and Appendini 2005), influencing land use by making maize production much less profitable.

Mexico's agricultural sector had been largely dependent on a protected market, high subsidies and low taxes. These policies became unsustainable as income from oil exports decreased (oil represented 72% of total exports until 1982). In 1985, the Mexican agricultural sector was restructured according to guidelines set by the World Bank and the International Monetary Fund (IMF) as a condition for further loans. The new economic strategy was oriented toward trade liberalization, reduction of restrictions on foreign investment, and reform of land tenure legislation (Bolay 1986). Mexico modernized its agricultural sector by moving from a predominantly subsistence-based agriculture to intensified production, and by introducing cash crops and commercial cattle ranching activities (Klepeis and Vance 2003). In 1986, Mexico signed the General Agreement on Tariffs and Trade (GATT). The impact was felt in the agricultural sector by 1990, when tariffs on most products were drastically lowered, subsidies on inputs were withdrawn or sharply reduced, and price controls were eliminated for all crops but maize and beans (Foley 1995). These reforms were continued under NAFTA, obligating Mexico to fully liberalize its agriculture, including maize and beans, over a 15-year period.

Economic liberalization was accompanied by legal reforms. Article 271 of the Constitution, which embodied the commitment to the rural poor since the end of the Mexican Revolution in 1917, was

---

[1] Article 27 of the Mexican Constitution declares that the wealth contained in the soil, the subsoil, the waters and seas of Mexico belongs to the Nation. The right to land ownership and to exploit the subsoil may therefore only be granted by the Nation. Land may also be expropriated whenever deemed necessary. This

amended in 1992 to 1) permit lands formerly held in usufruct under the ejido system to be bought and sold, 2) allow joint ventures between ejidos and private interests, and 3) end the distribution of land to peasant communities (Klepeis and Vance 2003).

Trade liberalization and the elimination of government subsidies under GATT and NAFTA has negatively affected producers in poor areas where markets and economic alternatives are limited (Deiniger and Minten 1999, Sánchez Reaza and Rodríguez Pose 2002). Results of the present study support the hypothesis that global factors increasingly influence land use change (Houghton 1994, Klooster 2003, Lambin et al. 2001, Lambin and Geist 2003, Mertz et al. 2005).

9.2 Driving forces at the national level

The case examples illustrate the importance of political/institutional driving forces. For instance, the **agricultural subsidy program**, PROCAMPO, is a significant source of income for farmers. It was thought that the financial support of PROCAMPO would make farmers more competitive in international markets and provide incentives for agricultural modernization (Klepeis and Vance 2003). In addition, crop-based subsidies were replaced by direct payments based on area cultivated, in order to benefit farmers who had previously produced too little to take advantage of price supports (Bonnis and Legg 1997). The PROCAMPO program was intended to foster social cohesion both by helping farmers adjust to the elimination of price supports and by including a broader range of agricultural producers (Klepeis and Vance 2003).

The primary goal of PROCAMPO was to support agricultural modernization and social welfare during the transition away from state intervention in the rural sector. A secondary goal was to decrease environmental degradation through the promotion of efficient land use (SAGARPA 2002). PROCAMPO was designed to slow environmental degradation, promote conservation and reforestation, reduce soil erosion and water pollution caused by the excessive use of non-organic pesticides, and promote sustainable development. Since the area and location eligible for support is fixed for the period 1994-2010, it was expected that the funds would be used to intensify production, thus decreasing pressure on the remaining forests (Klepeis and Vance 2003).

The addition of agave azul as a crop eligible for support in 1997 accelerated the recent expansion of cultivation of agave in the study area. Farmers decide to rent out their land because the annual rent,

---

article made it possible to control the activities of mining and oil companies, and to distribute the land of the large estates among the peasants.

though low, provides a higher income in combination with PROCAMPO subsidies than what could be obtained from cultivating maize. In addition, renting out land increases opportunities to seek off-farm employment or migrate to the US. Agricultural subsidies contributed to deforestation during the early 1990s,[2] when newly established pasture could be registered with PROCAMPO, encouraging clearing and cultivation of land in order to receive more subsidies. Subsidies for cattle ranching also caused deforestation, and led to the clearing of pasture in forested uplands, a process that has occurred in many Latin American countries (Lambin et al. 2001, Lambin et al. 2003, Masera et al. 1997). This study shows that PROCAMPO is a very important source of income for many households, constituting the second-largest source of income for 45% of all households in the case study sites. However, PROCAMPO subsidies are mainly used to cover subsistence needs and few households invest them to increase productivity.

Another important driving force in the Mexican context has been changes in the land tenure system. Since the land regularization program PROCEDE began in 1992, communal lands are being split up and distributed to individuals in all participating ejidos. As the situation in the ejido of Mezquitán indicates, the majority of landowners obtaining land titles to individual land plots of previously common land decide to deforest the newly acquired land for maize cultivation, pasture, or agave plantations, thereby converting forests and secondary vegetation to agriculture and pasture. In contrast, an absence of regularized land tenure, as in Chiquihuitlán and other communities (Gerritsen and Forster 2001), also contributes to deforestation as powerful families appropriate common land for private use, converting it to pasture. The absence of individual land titles combined with a high level of internal conflicts and corruption constitute important proximate causes for the ongoing deforestation in this area.

Bürgi et al. (2004) differentiated between intrinsic (e.g. community-level regulations) and extrinsic (e.g. legislation at the national or international level) driving forces. Extrinsic forces such as NAFTA and other trade liberalization agreements appear to be particularly strong influences. At the national level, changes in land tenure legislation and agricultural subsidy policies are proximate causes of extrinsic driving forces. However, intrinsic driving forces such as local environmental factors like topography, soil quality, and availability of irrigation water are equally important. The distinction between intrinsic and extrinsic (or endogenous and exogenous) factors is not easily made, as the interplay between organizational levels is complex. For example, community-level regulations are influenced by national legislation, just as national legislation is informed by

---

[2] The number of hectares eligible for PROCAMPO support is restricted to the area that was cultivated with a wide range of crops including maize, chili, and pasture in one of the three agricultural years prior to August 1994 (SAGARPA 2002).

international conventions. In this context the distinction between endogenous and exogenous is not particularly helpful. The decisive influences behind the observed land use changes are the multiple interactions between specific factors at different levels and not the predominance of one particular driving force functioning at a particular level.

9.3 Future trends in land use

The PROCAMPO program will end in 2010, at the same time that agrarian price supports are to be phased out under NAFTA (Klepeis and Vance 2003). What will be the effect on land use changes? Several scenarios are possible, in which the crucial factors will be: 1) The price of maize and other traditional crops, 2) The price of *agave azul*, 3) The cost of agricultural inputs, 4) The amount and variability of rainfall, 5) The effect of soil degradation on agave harvests, and 6) Whether or not households have the investment capital to diversify activities. The most likely scenario for the study area in 2010 is the following: The price for maize and other traditional crops will be even lower while the costs for agricultural inputs will have risen. Rainfall will continue to be low and irregular. The price for agave will have dropped further. International demand for tequila will continue to rise. Tequila companies will continue to rent land from individual farmers. Agave yield per hectare will have decreased between the first growing cycle (1996-2003) and the second growing cycle (2003-2010), but not to the extent that tequila companies leave the area. The association of independent agave producers will have grown and found a company that buys their agave harvest at a minimum guaranteed price. The possible consequences for the different groups of landowners classified according to income are outlined below:

1. Low-income farmers without investment capital who have the opportunity to rent their land to tequila companies will do so, because continuing to practice rainfed maize cultivation carries a high risk. Inputs and labor are too costly to merit the risky investment, due to the high probability of crop failure under conditions of late, low or variable rainfall. They would be producing maize at a net loss. The income from reverse leasing arrangements with the tequila companies will have decreased due to the lower price of agave, which lowers the amount of harvest income for the farmer. Also, the farmers will no longer be receiving PROCAMPO subsidies. Low-income farmers with rainfed land who do not have the option to rent out their land for agave cultivation nor the capital to cultivate agave themselves will be most affected by the end of PROCAMPO, as it was a crucial source of income for them and alternatives are few. Without alternatives, these farmers will continue to grow maize to cover subsistence needs. However, due to the high costs of chemical inputs, they will apply

low quantities or none and in consequence the harvests will tend to be very low. Some farmers might establish pasture on communal land to rent out to other farmers. This land use option represents a low risk. Pastures can be maintained with relatively low expenditures for labor and chemical inputs, while providing a good income. Migration to the US as a way to increase income will further gain in importance, but might be out of reach for farmers who cannot rent out their land and who do not have the necessary capital to migrate nor access to a network that can facilitate migration.

2. Middle-income farmers with rainfed land will either continue to rent out their land to tequila companies or independent agave cultivators, or if they have enough capital (e.g. remittances) they will become or continue as independent agave cultivators. Very few will choose to continue rainfed maize cultivation as profits are very low. Those who own cattle will continue maize cultivation on parts of their land for subsistence needs and to use the crop residue as fodder for their cattle. As many of the middle-income farmers have several sources of income, some might invest in cattle ranching and/or establishment of pastures. The disappearance of PROCAMPO subsidies will reduce their income but not drastically worsen their situation, as they have diversified income sources and therefore a better risk distribution than low-income farmers. Migration will gain in importance for the majority of this group. Many of the farmers of this group have the necessary capital allowing members of their family to migrate to the US, which they will do. On the other hand, some families might decide against migration because the need for additional income might not be strong enough.

3. High-income farmers will be the least affected by the end of the PROCAMPO program. They have several sources of income and do not depend on the subsidies from PROCAMPO. They will continue with independent agave cultivation and expand cattle ranching activities. They will not be incited to migrate to the US as a consequence of the end of PROCAMPO subsidies, as they have enough work and income in Mexico.

The consequences of this scenario imply: 1) Expansion of *agave azul* with continuous degradation of the soil, if management practices remain unchanged; 2) Continuation and expansion of cattle ranching activities and establishment of pasture; and 3) Increase in migration activities. This means that the land use change trend observed between 1990 and 2000 of increasing pasture and decreasing dry forest will continue.

The influence of national and global factors such as the decline of maize prices and changes in agricultural subsidy policies affect all farmers in Mexico. However, the local context of socio-economic, political and environmental factors determine to a large extent how people make land use decisions and react to economic, political or environmental changes. This makes generalization and an assessment of future trends at the national or even the state level difficult if not impossible.

9.4 The impact of migration and remittances

As illustrated by the case examples, the impact of remittances depends on the socio-economic, political and environmental context of the community. In communities with favorable conditions such as Mezquitán, there are more employment opportunities, easier access to markets, a stable land tenure situation and a low level of land tenure conflicts. Due to this positive context, families are more likely to have investment capital (remittances or other) and are also more likely to invest it in agriculture or house construction which leads to land use changes. There are also a number of wealthy families in the communities with less favorable conditions. They are in possession of investment capital for various reasons. Some accumulated it through migration, and others by belonging to the powerful families of the communities and thereby using a disproportionately large share of common resources, or through a combination of factors. The capital of private property owners stems from different sources, including either migration or successful off-farm activities. However, they all have in common favorable starting conditions, through belonging to wealthy families who have accumulated land, cattle and capital over decades.

In Chiquihuitlán 22% of all households receiving remittances invest it in a way that leads to land use changes, in El Jalocote this amount is 28% and in Mezquitán 41%. The total area affected by these land use changes is highly variable. It can consist of a minimal increase of urban area or may be tens of hectares of forested land converted into pasture. Furthermore, it has to be taken into account that these figures do not include households where remittances are used to cover current expenses, but free other sources of income to be invested in such a way as to change land use. It is not easy to answer the question of the relative importance of migration and remittances as proximate causes of land use change. It is certain that in some cases, remittances are the decisive factor driving land use changes. Remittances constitute an investment capital that is used for instance for land or cattle purchase, fencing material or the establishment of a cash crop such as agave, thereby changing land use very rapidly. Thus the original hypothesis of this research, that *migration is a driver of land use change*, can be accepted based on the role of remittances in driving land use change. With regard to labor, the results are ambiguous. As discussed in *chapter 8*, migration can drive land use change by encouraging the shift to low-labor land use systems, but these land use changes that require less labor can also drive migration.

Remittances are not invested in such a way as to gradually drive land use change but rather in a rapid, visible manner. The construction of houses constitutes an exception as some houses are built over several years, but the land area involved in the conversion of agricultural to urban land is not comparable to the large-scale conversions of forest to pasture or from maize to agave. Considering the limited number of households that are able to make such investments, in the study area remittances are only partially driving the extensive land use changes occurring in the region.

A number of issues related to the concept of remittance landscapes require further discussion. One of them is the conceptual distinction between migration and remittances. The study in the Philippines by McKay (2005) (see *chapter 3.3*) identified both remittances and migration, in particular the loss of female labour due to out-migration, as important drivers behind a land use change from subsistence rice production to commercial bean cultivation and qualified this transformation unequivocally as a remittance landscape. She argues that *"bean gardens can be read as remittance landscapes – they are sites for the investment of remittances and they are also a source of the capital outlay needed for overseas migration"* (McKay 2005). In her study, agricultural change was included as a land use change, and the site described as a remittance landscape because the change was driven by remittances and migration. The problem is one of terminology: *remittance landscape* clearly designates remittances as the main driver, whereas if the main factor behind a change is reduced labor availability due to out-migration, the emerging new type of landscape should logically be called *migration landscape* instead of *remittance landscape*. This leads to another problem, which is that remittances are inextricably linked to migration, and it is not easy to make a clear distinction between the two types. However, it seems clear that if the landscape transformation is driven by a change in labor availability due to migration, then the emerging landscape should not be included as a remittance landscape, as labor is the decisive factor behind the landscape transformation and not remittances. In addition, it seems coherent to require that the impact of remittances be apparent in transformations at the landscape level. Socio-economic impacts of the resulting remittance landscape should be treated separately. In contrast, a landscape described as *migration landscape* might include landscapes transformed by changes in labor availability as well as by the investment of remittances as remittances constitutes a sub-concept of migration.

Another issue is the question of the threshold: at which point does the impact of remittances on the landscape become such that the landscape can be designated a remittance landscape? This question has not been addressed in the studies by McKay (2003, 2005) that explicitly designate the emergence of "remittance landscapes." Existing studies often rely on qualitative assessments to identify remittance landscapes, such as that "much of the region" is converted, or that there is "significant" change. Jokisch (2002) found that in Highland Ecuador many migrant households abandoned agriculture and used remittances to purchase land, transforming the landscape from a predominantly agricultural to a peri-urban landscape. In that case, remittances were driving the conversion of *"much of the region into a peri-urban landscape of cultivated real estate."* However, he also does not specify a threshold for landscape transformation. What kind of impact qualifies as a "strong" or "significant" impact? Furthermore, few landscapes are static, hence the temporal dimension of transformation processes also needs to be considered. Quantitative indicators could specify that a certain percentage of total area is transformed, or that a certain percentage of landscape actors are participating in a particular practice of landscape transformation. However, even the appearance of single fields of a crop like *agave azul* can have a strong transformatory effect on the landscape through its environmental and visual impact. This striking transformation makes it tempting to conclude that this landscape change is significant, and should be defined as a remittance landscape **if** it is driven by the investment of remittances. The main question to be answered is the following: What are effective criteria against which the impact of remittances is measured? A combination of area-based and actor-based evaluation criteria seems appropriate in order to include quantitative as well as qualitative landscape transformations. An area-based criterion is efficient in evaluating large-scale transformations with regard to the percentage of the total surface of a landscape that is being transformed. This criterion might describe cases where a relatively small number of migrant families use remittances for major investments in cattle ranching, thereby deforesting large parts of a previously forested landscape, and ultimately transforming it into pasture. On the other hand, an actor-based criterion is efficient in evaluating processes where a majority of migrant families engage in certain transformation practices that do not necessarily transform large areas but bring with it significant qualitative changes at the landscape level. This criterion might describe cases where more than half of all migrant families use remittances for house construction, transforming an agricultural landscape with small adobe houses into a rural landscape with large, even ostentatious houses, giving the landscape a more urban character.

The definition of a universal criterion that takes into account the temporal dimension of transformation processes does not appear to be useful. Transformation processes linked to international migration can take place within a very short time or over decades. In Mexico for instance, international migration to the US started in the early 20th century. However, remittances began to really take on their present importance during the economic crisis in Mexico in the 1980s and 1990s. Therefore, in this research project it was decided to investigate the impact of remittances during the time period 1980-2004. In other countries, for instance in Central Asia, international migration and remittances are a relatively new phenomenon that only began in the 1990's. In this case, the determination of investigating a 15-year time period seems more appropriate. In view of the fact that there are "traditional" as well as "new" migrant sending countries as well as receiving countries, the temporal dimension needs to be determined according to the migration characteristics of the area under investigation.

Based on these reflections, the present study suggests the following definition and two criteria to evaluate whether a landscape can be described as a remittance landscape:

Definition:
Remittance landscapes are defined as *an emerging type of landscape driven by the investment of remittances*. Therefore, remittances must be the main driver of the landscape transformation that determines the evaluation criteria below.

Criteria:
Remittances are considered to be the main driver if a) more than 50% of the total area of an emerging land use is a result of the investment of remittances or b) if more than 50% of remittance-receiving landowners are participating in a particular practice of landscape transformation through the investment of remittances. This includes investments where remittances are used to cover certain expenses thereby freeing other sources of income that are invested in a way that leads to land use changes.

However, it is important to conduct additional research in other areas in remittance-receiving countries - such as for instance China, the Philippines and Morocco - to compare investment characteristics and possibly further refine the definition and evaluation criteria of a remittance landscape.

Agave azul has a strong visual impact on the landscape. The expansion of agave in the region resembles a large blue wave invading the valleys and creeping up the hills. Photos by author (March 2003).

9.5 Recommendations

1. **Agricultural subsidy policy**

   Phasing out of PROCAMPO subsidies should be accompanied by assistance to low-income farmers. For low-income farmers, PROCAMPO subsidies are essentially a welfare payment. The PROCAMPO payments are not sufficient to induce changes in farming strategies and increase productivity. The scarcity of water and the need for costly chemical inputs represent obstacles that are beyond the scope of the individual farmer to resolve. There is no ready solution, as better access to credit and markets will not be enough to overcome the limiting factors. The vulnerability of low-income farmers with very few alternatives requires that they receive continued assistance beyond the PROCAMPO program.

2. **Sustainable land management**

   Private industry has a central role in the transition from semi-subsistence to commercially oriented, agroindustrial production, especially in areas where low-income farmers have little capital to invest in new production systems. With regard to *agave azul* cultivation three issues need to be addressed: 1) Better management practices for *agave azul* cultivation need to be developed and implemented in order to reduce the risk of soil erosion and chemical soil degradation, 2) A minimum price for agave should be guaranteed by the tequila companies buying agave, and 3) Employment of farmers as agricultural workers by the tequila companies requires more adequate regulations, for instance with regard to salaries and the provision of appropriate protection when handling chemicals. Improvement of agave cultivation practices requires collaboration between private industry, the independent agave grower association, the communities growing agave or renting out land to tequila companies, the municipal government, state agencies and scientists. However, effective collaboration and negotiation with tequila companies in the region will be challenging due to unequal power relations between the different actors. Ongoing efforts include a research project by the University of Guadalajara in the neighboring municipality of Tonaya on the social, economic and ecological sustainability of *agave azul* cultivation. This kind of activity will hopefully pave the way for expanded collaboration.

The ongoing trend of deforestation also raises several issues in resource management. The production of charcoal needs to be better controlled by reducing the number of permits and/or through an increased presence of forestry officials enforcing the restriction on producing charcoal only in the authorized forest areas. In addition, the evolution of cattle ranching and pasture establishment needs to be closely followed. With the further decline of maize prices and the increase of costs for chemical inputs, it is likely that a growing number of farmers will turn to cattle ranching activities. Those farmers who have the necessary capital may increasingly invest it in cattle and pastures, and those without capital are likely to establish and rent out pasture as a low-cost, low-risk land use option that provides a good income.

Finally, efforts should be undertaken to reduce land tenure conflicts in a number of communities, especially in *comunidades agrarias* where the land regularization program PROCEDE has not been conducted. This requires interdisciplinary teams that are skilled in conflict resolution. In view of the complexity of such conflicts, some of which have endured over several decades, this will undoubtedly be an extremely difficult task and require a lot of time and effort. However, as long as these issues are not resolved, any program linked to sustainable land management will have limited chances of success.

9.6 Further research needs

The results of this study lead to a number of further conceptual and topical research questions:

1. **Conceptual research needs**:
    a) **Further development of the concept of remittance landscape.** In view of the global increase of migration and the growing importance of remittances in developing countries, the concept of remittance landscape deserves further

reflection. Relevant questions include: What type of remittance landscape is likely to emerge in particular socio-economic, political and environmental contexts? What are the socio-economic and environmental impacts of the transformation of landscapes into remittance landscapes for different groups of stakeholders? How can these transformation processes be linked to policy in order to increase sustainability?

b) **Up-scaling of results on land use change.** Identifying the causes of land use change requires an understanding of how people make land use decisions and how various factors interact in specific contexts to influence this decision-making. Even though the concepts of proximate causes and underlying driving forces allow for a certain amount of generalization, due to the importance of local context, up-scaling of results from the local to regional and national levels is quite difficult. The differences in land use dynamics between communities in this study demonstrate the decisive influence of local context on land use changes. The inherently local nature of household-based land use change makes the development of a general land use change theory elusive. At the same time, if land use change research is to constructively contribute to policy development, a certain level of generalization is necessary.

2. **Topical research needs** :

   a) Quantification of soil erosion and long-term monitoring of soil degradation on *agave azul* fields;

   b) Development of more sustainable management practices for the cultivation of *agave azul*;

   c) Evaluation of the impact of the end of the federal agricultural subsidy program PROCAMPO on deforestation trends.

10. Conclusions

This study explored land use change processes in a particular area in Mexico. It has expanded knowledge of how various factors interact in a specific context to influence land use decision-making. It has also contributed to the understanding of cross-scalar dynamics between macro-level economic phenomena such as NAFTA and local-level decision-making, and how these interactions drive processes of land use change. Furthermore, the study links international trade agreements and governmental policy initiatives such as PROCAMPO to local environmental impacts such as the soil degradation associated with *agave azul* cultivation.

In this concluding chapter, a summary of the results is given with regard to the four initial research objectives of this study.

1. Assessment of land use changes in the municipality of Autlán from 1990 to 2000.
Land use changes observed in the study area between 1990 and 2000 include a slight increase of agricultural land (2%), urban land cover (0.5%) and pine-oak forest (0.7%). Over the same period, pasture increased by 18% while dry forest decreased by 10%. Both of these trends are close to the official averages of changes in land use type for the state of Jalisco. Rapid and extensive land use change is occurring on rainfed agricultural land, as maize cultivation is converted to the cultivation of *agave azul* for the production of tequila. The first plantations of *agave azul* were established in 1996, and by 2002 agave azul was planted on 33% of all rainfed agricultural land in the municipality, and in some cases on the entire agricultural land of some communities. 84% of owners of rainfed land included in the survey had changed land use from maize to agave cultivation during this time period.

2. Assessment of the driving forces and dynamics underlying those changes.
Economic driving forces are central because land use changes are largely motivated by market prices for different products and the costs of agricultural inputs, which in turn are influenced by international markets and trade agreements. However, these are not sufficient to drive land use change, and interaction with other factors such as policies and the institutional context is equally important. Lambin et al. (2001) proposed that the main underlying driving forces of land use change are economic opportunities mitigated by institutional factors. The results of the present study would support this hypothesis, but include environmental factors as an equally important driving force. The irregularity of rainfall and absence of irrigation water are key factors driving land use change in the study area. At the same time, where irrigation water is available, it is an equally important factor

stabilizing current land use systems. The large majority of landowners emphasize that if rainfall was regular and sufficient, or irrigation water available, they would not have changed land use. Furthermore, topography and soil quality play an essential role as environmental factors that are decisive as to whether or not a certain land use change will occur. The case study in El Jalocote has shown that small and steep land plots that are not easily accessible are not attractive to tequila companies. That environmental characteristic, in combination with the presence of irrigation water and the economic profits to be obtained through irrigated agriculture, is a main reason that land use change to agave has not occurred in this area. This illustrates how biophysical drivers are as important as socio-economic drivers.

The main driving forces identified in the study area are economic, political/institutional and environmental. Demographic factors play a role with regard to labor availability but are not as important as other factors. Environmental factors are crucial driving forces for land use change on agricultural land but less so for land use change in the hills (see **Figure 10-01**). Technology and culture appear to be less important in both land use contexts.

Figure 10-01 **Overview of proximate causes and driving forces in study area**

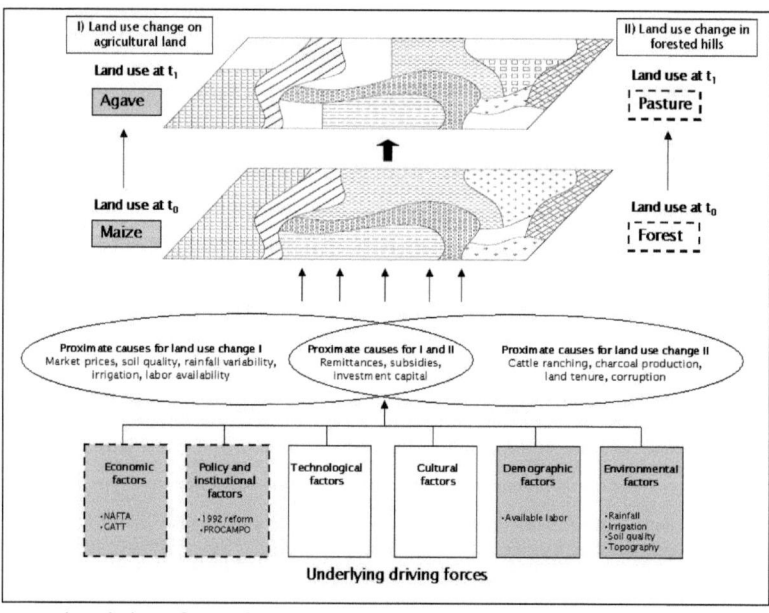

Legend: Underlying forces driving land use change from maize to agave are marked in dark grey. Underlying forces driving land use change from forest to pasture are marked with dashed lines. Diagram by author.

3. Appraisal of environmental implications of land use changes and future trends

Agave azul cultivation was assessed with regard to soil erosion. The evaluation of soil erosion risk level was based on the location of agave fields with regard to slope. Results show that 72% of all agave fields are located on slopes above 2%, thus requiring conservation measures such as contour cultivation or strips of grass and/or trees along the contours. These conservation measures are extremely rare. On the vast majority of agave fields, plants are planted in straight rows perpendicular to the slope. Due to low ground cover during the first 4-5 years of agave cultivation, the risk of soil erosion is high. It is estimated that 28% of fields will show signs of slight erosion, 17% signs of slight to moderate erosion, 30% signs of moderate erosion, 16% signs of moderate to severe erosion, while 9% of fields are likely to show evidence of severe to very severe erosion. Furthermore, application rates of pesticides, herbicides and fertilizer for *agave azul* are higher than for maize cultivation, suggesting a higher chemical degradation of the soil.

Harris (2000) defined environmental sustainability as follows: *"An environmentally sustainable system must maintain a stable resource base, avoiding over-exploitation of renewable resource systems or environmental sink functions, and depleting non-renewable resources only to the extent that investment is made in adequate substitutes. This includes maintenance of biodiversity, atmospheric stability, and other ecosystem functions not ordinarily classed as economic resources."* The land use change from maize to *agave azul* and from dry forest to pasture cannot be considered sustainable according to this definition nor according to the principles of sustainable land management (Hurni 2000). Neither of these land use changes maintain a stable resource base as forest disappears, soil is degraded and biodiversity declines. Social and economic sustainability were not systematically evaluated in this study but according to the definitions by Harris,[3] (2000) the land use change to *agave azul* and pasture do not correspond to economically or socially sustainable land use systems.

4. Investigation of the impact of migration and remittances on land use changes

Labor migration is a significant livelihood strategy in the study area, enhancing the financial resources and security of migrant households by diversifying income and spreading risk. On average 50% of all households in the survey have, or had a family member in migration in the US

---

[3] An **economically sustainable system** must be able to produce goods and services on a continuing basis, to maintain manageable levels of government and external debt, and to avoid extreme sectoral imbalances which damage agricultural or industrial production (Harris 2000).
A **socially sustainable system** must achieve distributional equity, adequate provision of social services including health and education, gender equity, and political accountability and participation (Harris 2000).

between 1980 and 2004. The remittances sent back by these family members constitute a crucial source of income as they represent on average 45% of total household earnings.

The study by Portner (2005) identified several differences in land use strategies between migrant and non-migrant households. For instance, migrant households show more innovative land use strategies than non-migrant households. It showed that labor and investment capital are the two decisive factors on which landowners base their land use strategies. An important difference was that migrant households have the necessary resources to complete agricultural tasks on time and thereby achieve a better harvest. However, environmental factors such as topography, soil quality and the availability of irrigation water proved to be more important influences on land use than whether the household had migration experience or not. Nonetheless, 22% of migrant households invested remittances in ways that led to land use changes, mainly by investing in cattle ranching and pasture.

Remittances did play a role in processes of land use change, with three main findings from this study: First, remittances drive land use changes as a proximate cause for economic factors thus the original hypothesis of this research, that *migration is a driver of land use change*, can be accepted. Second, even though on average 30% of migrant households invest remittances in a way that leads to land use changes, the importance of remittances is relatively low compared to other proximate causes such as market prices and the investment capital procured by secure off-farm employment. Third, in cases where remittances lead to land use changes, these changes occur rapidly.

Migration and remittances increase in extent and importance at the global level. Therefore, tools are required to analyse their impact not only in destination countries but also in the sending communities. The concept of remittance landscape developed by the researcher and defined as *an emerging type of landscape driven by the investment of remittances*, has proved useful for analysing the impact of remittances on land use changes. Based on the results of this study, four main conclusions can be drawn :

1) The criteria determining the threshold at which a changing landscape becomes a remittance landscape were challenging to determine. Based on the definition of remittance landscape, remittances must be the main driver behind the landscape transformation. Consequently, two evaluation criteria have been determined: Remittances are considered to be the main driver if a) more than 50% of the total area of an emerging land use is a result of the investment of remittances, **or** b) if more than 50% of remittance-receiving land owners are participating in

a particular practice of landscape transformation through the investment of remittances. This includes investments where remittances are used to cover certain expenses thereby freeing other sources of income that are invested in a way that leads to land use changes.

2) Landscapes where the investment of remittances leads to a change of land use from subsistence to cash crop cultivation should be included as a potential type of remittance landscape, even though the basic type of the landscape (agricultural) remains unchanged.

3) In landscapes where the reduced availability of labor due to out-migration leads to land use changes, the transformed landscape should not be designated as a remittance landscape since labor is the decisive driver behind the transformations and not the investment of remittances.

4) In conclusion, in the study area where this research was conducted, the investment of remittances is visible as changes in the landscape. However, not all of these changes can be qualified as remittances landscapes as defined under 1) because in many areas remittances are only partially driving the extensive land use changes occuring in the region.

References

Adams R, Page J. 2005. Do International Migration and Remittances Reduce Poverty in Developing Countries? World Development 33(10):1645-1669.

Agarwal C, Green GM, Grove JM, Evans TP, Schweik CM. 2002. A review and assessment of land-use change models: dynamics of space, time and human choice: United States Department of Agriculture (Northeastern Research Station), Indiana University, (Center for the Study of Institutions, Population, and Environmental Change). 29 p.

Alfieri A, Havinga I. 2005. Definition of Universe for the Framework on the Movement of Persons. United Nations Technical Subgroup on the Movement of Natural Persons (TSG). Report nr BOPCOM-05/9A.

Appendini K. 2002. Land regularization and conflict resolution: The case of Mexico. Rome: FAO. 54 p.

Artes de México. 1999. El Tequila. Arte tradicional de México. Artes de México 27.

Aymer P. 1997. Uprooted women: migrant domestics in the Caribbean. Connecticut: Praeger.

Bailey AJ. 2001. Turning transnational: notes on the theorisation of international migration. International Journal of Population Geography 7(6):413-428.

Banerjee SK, Jayachandran V, Roy TK. 2002. Has emigration influenced Kerala's living standards? A micro level investigation. Economic and Political Weekly:1755-1765.

Barrett CB, Barbier EB, Reardon T. 2001. Agroindustrialization, globalization, and international development: the environmental implications. Environment and Development Economics 6:419-433.

Basch L, Schiller NG, Blanc-Szanton C. 1994. Nations unbound: Transnational projects, post-colonial predicaments, and de-territorialized nation-states: Gordon and Breach, Langhorne, PA.

Basok T. 2003. Mexican Seasonal Migration to Canada and Development: A Community-based Comparison. International Migration 41(2):3-26.

Baulies X, Szejwach G. 1998. LUCC data requirements workshop - survey of needs, gaps and priorities on data for land-use/land-cover change research. Institut Cartogràfic de Catalunya.

Bean FD, Corona R, Tuirán R, Woodrow-Lafield KA. 1998. The quantification of migration between Mexico and the United States. In: Migration between Mexico and the United States: Binational Study. Mexico City and Washington, D.C.: Mexican Ministry of Foreign Affairs and U.S. Commission on Immigration Reform.

Bebbington AJ, Batterbury SPJ. 2001. Transnational livelihoods and landscapes: political ecologies of globalization. Ecumene 8(4):369-380.

Bewket W. 2002. Land cover dynamics since the 1950's in Chemoga watershed, Blue Nile basin, Ethiopia. Mountain Research and Development 22(3):263-269.

Bichsel C, Hostettler S, Strasser B. 2005. "Should I buy a cow or a TV?". Reflections on the conceptual framework of the NCCR North-South based on a comparative study of international labour migration in Mexico, India and Kyrgyzstan. Berne: NCCR North-South dialogue. 60 p.

Blaikie P. 1985. The political economy of soil erosion in developing countries. Essex, UK: Longman.

Blaikie P. 1994. Political ecology in the 1990s. An evolving view of nature and society; Michigan State University.

Blaikie P. 1995. Changing environments or changing views: A political ecology of developing countries. Geography 80:203-214.

Blaikie P. 1999. A review of political ecology - Issues, epistemologies and analytical narratives. Zeitschrift für Wirtschaftsgeographie 43(3-4):131-147.

Blaikie P, Brookfield H. 1987. Land degradation and society. London: Methuen.

Blanco Barbosa J. 2004. Interview with José Blanco Barbosa, landowner in Chiquihuitlán. Chiquihuitlán (29 March 2004).

Bonnis G, Legg W. 1997. The opening of Mexican agriculture. OECD observer 206:35-38.

Bolay J-C. 1986. Les migrants dans la ville. Un cas mexicain: Toluca et sa région. Berne: Editions Peter Lang SA. 229 p.

Bolay J-C. 2004. World globalisation, sustainable development and scientific cooperation. International Journal of Sustainable Development 7:99-120.

Bolay J-C, Hostettler S, Pleyan CG. 2004. JACS Central America and the Caribbean. Key challenges of sustainable development and research priorities: Social practices as driving forces of change. In: Hurni H, Wiesmann U, Schertenleib R, editors. Research for mitigating syndromes of global change. A transdisciplinary appraisal of selected regions of the world to prepare development-oriented research partnerships. Berne: University of Berne: Geographica Bernensia. p 468.

Boserup E. 1965. The conditions of agricultural growth. London: Earthscan Publications Ltd.

Bowen S. 2004. The road to Margaritaville: Expansion of agave cultivation and power dynamics in southern Jalisco, Mexico [MSc thesis]. Wisconsin: University of Wisconsin-Madison (unpublished). 125 p.

Briassoulis H. 1999. Analysis of land use change: Theoretical and modeling approaches. Regional Research Institute. The Web Book of Regional Science: West Virginia University.
www.rri.wvu.edu/WebBook/Briassoulis/contents.htm

Bruinsma J, editor. 2003. World Agriculture: Towards 2015/2030. An FAO perspective. London: Earthscan Publications. 520 p.

Bryant RL, Bailey S. 1997. Third World Political Ecology. London: Routledge.

Bürgi M, Hersperger AM, Schneeberger N. 2004. Driving forces of landscape change - current and new directions. Landscape Ecology 19:857-868.

Bürgi M, Turner MG. 2002. Factors and Processes Shaping Land Cover and Land Cover Changes Along the Wisconsin River. Ecosystems 5(2):184-201.

Cassels S, Curran SR, Kramer R. 2005. Do migrants degrade coastal environments? Migration, natural resource extraction and poverty in North Sulawesi, Indonesia. Human Ecology 33(3):329-363.

Chambers R. 1983. Putting the Last First. Harlow, UK: Longman.

Chapman M, Prothero M. 1985. Circulation in population movement: substance and concepts from the Melanesian Case. London: Routledge and Kegan Paul.

CONAPO. 2003. Proyecciones de la población municipal 1995-2010. www.conapo.gob.mx

Conway D, Lorah P. 1995. Environmental protection policies in Caribbean small Islands: Some St. Lucian Examples. Caribbean Geography 6(1):16-27.

Creswell JW. 2003. Research design: Qualitative, quantitative, and mixed method approaches. Thousand Oaks: Sage Publications. 246 p.

Cruz Mercado J. 2004. Interview with José Cruz Mercado. *Comisariado* of the *ejido* Mezquitán. Mezquitán (4 April 2004).

Curran S. 2001. Migration, social capital, and the environment: Considering migrant selectivity and networks in relation to coastal ecosystems. Working Paper no. 01-02. Princeton: The Center for Migration and Development. Princeton University.

David R. 1995. Changing places: Women, resource management and migration in the Sahel. London: SOS Sahel.

de Haan A. 1999. Livelihoods and poverty: The role of migration - A critical review of the migration literature. The Journal of Development Studies 36(2):1-47.

de Haan A, Brock K, Coulibaly N. 2002. Migration, Livelihoods and Institutions: Contrasting Patterns of Migration in Mali. The Journal of Development Studies 38(5):37-58.

de Haan A, Rogaly B. 2002. Introduction: Migrant Workers and Their Role in Rural Change. The Journal of Development Studies 38(5):1-14.

de Haan LJ. 2000. Globalization, Localization and Sustainable Livelihood. Sociologia Ruralis 40(3):339-365.

de Haan LJ, Zoomers A. 2005. Exploring the Frontier of Livelihoods Research. Development and Change 36(1):27-47.

de Haas H. 2005. International migration, remittances and development: myths and facts. Third World Quarterly 26(8):1269-1284.

Deininger KW, Minten B. 1999. Poverty, policies, and deforestation: The case of Mexico. Economic Development and Cultural Change 47(5):313-344.

Delgado R, Rodríguez H. 2001. The emergence of collective migrants and their role in Mexico's local and regional development. Canadian Journal of Development Studies 22(3):747-764.

Derman B, Ferguson A. 2000. The value of water: political ecology and water reform in southern Africa. Panel on political ecology for the annual meeting of the American Anthropological Association. San Francisco: Department of Anthropology, Michigan State University.

Deshingkar P, Start D. 2003. Seasonal migration for livelihoods in India: Coping, accumulation and exclusion. Working paper nr 220. London: Overseas Development Institute.

Durand J. 1994. Más allá de la línea: Patrones migratorios entre México y Estados Unidos. México. D.F: Consejo Nacional para la Cultura y las Artes.

Durand J, Massey DS. 1992. Mexican migration to the United States: A critical review. Latin American Research Review 27(2):3-42.
Durand J, Massey DS. 2003. Clandestinos. Migración México-Estados Unidos en los albores del siglo XXI. Zacatecas: Universidad Nacional Autónoma de Zacatecas.
Durand J, Parrado EA, Massey DS. 1996. Migradollars and development: a reconsideration of the Mexican case. International Migration Review 30(2):423-444.
Eakin H, Appendini K. 2005 Subsistence maize production and maize liberalization in Mexico. IHDP Update 01/2005:4-6.
Ediger L. 2006. Afforestation and household resource allocation in Yunnan, China: Implications for landscape patterns and rural livelihoods. [PhD thesis] Athens: University of Georgia (unpublished). 142 p.
Ediger L, Huafang C. 2006. Upland China in transition. Mountain Research and Development 26(3):220-226.
Ellerman D. 2005. Labour migration: a developmental path or a low-level trap? Development in Practice 15(5):617-630.
Ellis F. 1998. Household Strategies and Rural Livelihood Diversification. The Journal of Development Studies 35(1):1-38.
Ellis F, Biggs S. 2001. Evolving themes in rural development 1950s-2000s. Development Policy Review 19(4):437-448.
Evans TP, Ostrom E, C.Gibson. 2003. Scaling issues in the social sciences. In: Rotmans J, Rothman DS, editors. Scaling in Integrated Assessment. Lisse, The Netherlands: Swets and Zeitlinger BV.
Fairhead J, Leach M. 1996. Misreading the African landscape: Society and ecology in a forest-savanna mosaic. Cambridge: Cambridge University Press.
FAO. 1990. Guidelines for soil description. Rome: FAO.
FAO. 1990a. Evaluación de los Recursos Forestales de 1990: Informe de México. Mexico City: Food and Agriculture Organization.
FAO. 1995. Global and national soils and terrain digital databases (SOTER). Procedures manual. Rome: FAO Land and Water Development Div.; UNEP, Geneva (Switzerland); International Society of Soil Science, Wageningen (Netherlands). 125 p. www4.fao.org/faobib/index.html
FAO. 2000a. Global Agro-Ecological Zones. Agro-edaphic suitability analysis. www.fao.org/ag/agl/agll/gaez/method/c_45.htm#4_4_2
FAO. 2000b. Watershed-L. The electronic workshop on land-water linkages in rural watersheds.: FAO. Electronic workshop.
www.fao.org/ag/AGL/watershed/watershed/en/mainen/index.stm
FAO. 2001. Guidelines for the qualitative assessment of land resources and degradation. Rome: FAO and ISRIC. 40 p.
Faux F. 2006. L'exode des Mexicains aux Etats-Unis vide les campagnes mais remplit les poches. Le Temps, 1 July 2006;4.
Flora C. 2001. Interactions between agroecosystems and rural communities. Series: Advances in Agroecology: CRC Press.

Flores Preciado EM, Zamora Durán JJ. 2003. Análisis socioambiental de la expansión del cultivo de *agave azul* (*Agave tequilana* WEBER) en los municipios de Autlán de Navarro y Tuxcacuesco, Jalisco [BSc thesis]. Autlán: University of Guadalajara (unpublished). 125 p.

Foley MW. 1995. Privatizing the countryside: The Mexican peasant movement and neoliberal reform. Latin American Perspectives 22:59-76.

Forsyth T. 2003. Critical political ecology: the politics of environmental science. London and New York: Routledge.

Fox J, Rambo T, Donovan D, Cuc LT, Giambelluca T, Ziegler A, Plondke D, Vien TD, Leisz S, Truong DM. 2003. Linking household and remotely sensed data for understanding forest fragmentation in northern Vietnam. In: Fox J, Rindfuss RR, Walsh SJ, Mishra V, editors. Linking household and remotely sensed data for understanding forest fragmentation in northern Vietnam. Dordrecht, Netherlands: Kluwer Academic Publishers. p 201-221.

Francis & Taylor. 2006. Journal of Land Use Science: Call for papers. Taylor & Francis. www.tandf.co.uk/journals/titles/1747423X.asp

Gammeltoft P. 2002. Remittances and Other Financial Flows to Developing Countries. International Migration 40(5):181-211.

García y Griego M. 1998. Responses to migration: The Bracero program. Migration Between Mexico and the United States: Binational Study. Mexico City and Washington, D.C.: Mexican Ministry of Foreign Affairs and U.S. Commission on Immigration Reform. p 1215-1221.

García-Zamora R. 2006. A better quality of life? Insights: id21. www.id21.org

Gautam AP, Webb EL, Eiumnoh A. 2002. GIS Assessment of land use / land cover changes associated with community forestry implementation in the middle hills of Nepal. Mountain Research and Development 22(1):63-69.

Geist H. 2006. Our Earth's Changing Land. An Encyclopedia of Land-Use and Land-Cover Change. Westport, Connecticut, London: Greenwood Press.

Geist HJ, Lambin EF. 2002. Proximate causes and underlying forces of tropical deforestation. BioScience 52(2):143-150.

Gerritsen P. 2002. Diversity at Stake. A farmers' perspective on biodiversity and conservation in western Mexico. Wageningen Studies on Heterogeneity and relocalisation. University of Wageningen. 286 p.

Gerritsen P, Forster N. 2001. Conflict over natural resources and conservation in the indigenous community of Cuzalapa, Western Mexico. In: Zoomers A, editor. Land and sustainable livelihood in Latin America. Koninkliijk Instituut Voor de Tropen: Vervuert Verlag. p 139-155.

Gibson CC, E. Ostrom, Ahn TK. 2000. The concept of scale and the human dimensions of global change: a survey. Ecological Economics 32:217-239.

Glick-Schiller N, Basch L, Szanton-Blanc C. 1992. Towards a transnational perspective on migration: Race, class, ethnicity, and nationalism reconsidered. New York: The New York Academy of Sciences. 259 p.

Global Land Project. 2005. Science plan and implementation strategy. Stockholm. www.glp.colostate.edu

Glytsos NP. 2002. The Role of Migrant Remittances in Development: Evidence from Mediterranean Countries. International Migration 40(1):5-26.

Goldring L. 2004. Family and Collective Remittances to Mexico: A Multidimensional Typology. Development and Change 35(4):799-840.

Goméz Garcia M. 2003. Interview with Ing. Martin Goméz Garcia. Director of the Sierra de Manantlán Biosphere Reserve. Autlán (12 February 2003).

Gonzáles A. 2002. Blue agave peasant producers and the tequila industry in Jalisco, Mexico. Oxford: University of Oxford. 47 p.

Greenberg JB, Park TK. 1994. Political Ecology. Journal of Political Ecology 1:1-12.

Gundel J. 2002. The Migration-Development Nexus: Somalia Case Study. International Migration 40(5):255-281.

Guyer JI, Lambin EF. 1993. Land use in an urban hinterland: ethnography and remote sensing in the study of African intensification. American Anthroplogist 95(4):839-859.

Hansen TS. 2005. Land use and land cover changes in the Niah catchment, Sarawak, Malaysia: Linking driving forces, land use, and river water quality. Copenhagen: University of Copenhagen. 313 p.

Harris JM. 2000. Basic Principles of Sustainable Development. Working Paper 00-04 Medford, MA: Tufts University, Global Development and Environment Institute. http://ase.tufts.edu/gdae/publications/working_papers/Sustainable%20Development.PDF

Hostettler S. 2006. Remittances landscapes. In: Geist H, editor. Our Earth's Changing Land. An Encyclopedia of Land-Use and Land-Cover Change. Westport Connecticut, London: Greenwood Press. p 503-506.

Houghton RA, Boone RD, Melillo JM, Palm CA, Woodwell GM, Myers N, Moore B, Skole DL. 1985. Net flux of carbon dioxide from tropical forests in 1980. Nature 316:617-620.

Houghton RA. 1994. The worldwide extent of land-use change. BioScience 44(5):305-313.

Hurni H. 1998. A Multi-Level Stakeholder Approach to Sustainable Land Management. Advances in GeoEcology 31:827-836.

Hurni H. 2000. Assessing Sustainable Land Management (SLM). Agriculture, Ecosystems and Environment 81:83-92.

Hurni H, Wiesmann U, Schertenleib R, editors. 2004. Research for Mitigating Syndromes of Global Change. A Transdisciplinary Appraisal of Selected Regions of the World to Prepare Development-Oriented Research Partnerships. Berne: Geographica Bernensia. 468 p.

Hurni H, editor. 2002. A world soils agenda. Discussing international actions for the sustainable use of soils. Prepared with the support of an international group of specialists of the IASUS working group of the International Union of Soil Sciences (IUSS). Berne: Centre for Development and Environment. 63 p.

IIED. 2003. Contribution by the International Institute for Environment and Development and its partners to the UK International Development Committee's inquiry on Migration and Development. International Institute for Environment and Development. www.iied.org/docs/urban/IDC_submission_migration.pdf

IIED. 2003. Rural-urban transformations and the links between urban and rural development. Environment & Urbanization 15(1):1-6.

INEGI. 1975. V Censos agricola-ganadero y ejidal 1970. Aguascalientes: Direccion general de Estadistica Jalisco.

INEGI. 1991. VII Censo Ejidal 1991. Ejidos y comunidades agrarias, superficie y actividad principal. http://sc.inegi.gob.mx/simbad/index.jsp

INEGI. 1993. Anuario estadístico del estado de Jalisco. Aguascalientes: Gobierno del estado de Jalisco.

INEGI. 1994. Anuario estadístico del estado de Jalisco. Aguascalientes: Gobierno del estado de Jalisco.

INEGI. 1995. Anuario estadístico del estado de Jalisco. Aguascalientes: Gobierno del estado de Jalisco.

INEGI. 1996. Anuario estadístico del estado de Jalisco. Aguascalientes: Gobierno del estado de Jalisco.

INEGI. 1997. Anuario estadístico del estado de Jalisco. Aguascalientes: Gobierno del estado de Jalisco.

INEGI. 1998. Anuario estadístico del estado de Jalisco. Aguascalientes: Gobierno del estado de Jalisco.

INEGI. 1999. Anuario estadístico del estado de Jalisco. Aguascalientes: Gobierno del estado de Jalisco.

INEGI. 2000. Anuario estadístico de Jalisco. Aguascalientes: Gobierno del estado de Jalisco.

INEGI. 2001. Resultados del VIII Censo Ejidal 2001 Guadalajara: INEGI. www.inegi.gob.mx/prod_serv/contenidos/espanol/biblioteca/Default.asp?accion=1&upc=702825000383

INEGI. 2005. Sistema para la consulta del cuaderno estadistico de Autlán de Navarro, Jalisco Edicion 2003. www.inegi.gob.mx/est/contenidos/espanol/sistemas/cem03/estatal/jal/m015/index.htm

Islam MD. 1991. Labour migration and development: A case study of a rural community in Bangladesh. Bangladesh Journal of Political Economy 11(2):35-47.

Jardel E, Santana EC, Graf S. 1996. The Sierra de Manantlán Biosphere Reserve: conservation and sustainable development. Parks 6(1):14-22.

Jennings A, Clarke M. 2005. The development impact of remittances to Nicaragua. Development in Practice 15(5):685-692.

Johnson NC, Malk AJ, Szaro RC, Sexton WT. 1999. Ecological stewardship - a common reference for ecosystem management. UK: Elsevier Science Ltd.

Jokisch B. 2002. Migration and Agricultural Change: The Case of Smallholder Agriculture in Highland Ecuador Human Ecology 30(4):523-550.

Jones RC. 1995. Ambivalent journey: US migration and economic mobility in North-Central Mexico. Tucson: University of Arizona Press.

Jones H, Pardthaisong T. 1999. The impact of overseas labour migration on rural Thailand: Regional, community and individual dimensions. Journal of Rural Studies 15(1):35-47.

Kanaiaupuni SM. 2000. Sustaining families and communities: Nonmigrant women and Mexico-U.S. migration processes. Madison, Wisconsin: University of Wisconsin-Madison.

Kannan KP. 2005. Kerala's turnaround in growth. Role of social development, remittances and reform. Economic and Political Weekly (5 February 2005):548-554.

Kapur D. 2004. Remittances - the new development mantra?: Harvard University and Center for Global Development. 20 p.

Kaufmann V, Bergman MM, Joye D. 2004. Motility: Mobility as Capital. International Journal of Urban and Regional Research 28(4):745-756.

Klepeis P, Vance C. 2003. Neoliberal Policy and Deforestation in Southeastern Mexico: An Assessment of the PROCAMPO Program. Economic Geography 79(3):221-240.

Klooster D. 2003. Forest transition in Mexico: Institutions and forests in a globalized countryside. The Professional Geographer 55(2):227-237.

Lambin EF. 1993. Spatial scales and desertification. Desertification Control Bulletin 23. UNEP. p 20-23.

Lambin EF, Baulies X, Bockstael N, Fischer G, Krug T, Leemans R, Moran EF, Rindfuss Y, Sato Y, Skole D and others. 1999. Land-use and land-cover change (LUCC) Implementation Strategy. Stockholm and Bonn: IGBP, IHDP.

Lambin EF, Geist HJ. 2003. Regional differences in tropical deforestation. Environment 45(6):22-36.

Lambin EF, Geist HJ, Lepers E. 2003. Dynamics of land-use and land-cover change in tropical regions. Annual Review of Environment and Resources 28:205-241.

Lambin EF, Turner BL, Geist HJ, Agbola SB, Angelsen A, Bruce JW, Coomes OT, Dirzo R, Fischer G, Folke C and others. 2001. The causes of land-use and land-cover change: moving beyond the myths. Global Environmental Change 11:261-269.

Landolt P. 2001. Salvadorian economic transnationalism: embedded strategies for household maintenance, immigrant incorporation, and entrepreneurial expansion. Global Networks 1(3):217-242.

Laumann G. 2006. "Handing over" - From LUCC to GLP. IHDP Update:25.

León-Ledesma M, Piracha M. 2004. International Migration and the Role of Remittances in Eastern Europe. International Migration 42(4):65-83.

Lepers E, Lambin EF, Janetos AC, Defries R, Achard F, Ramankutty N, Scholes RJ. 2005. A synthesis of information on rapid land-cover change for the period 1981-2000 BioScience 55(2):115-124.

Levitt P. 1998. Social remittances: Migration driven local-level forms of cultural diffusion. International Migration Review 32(4):926-948.

Levitt P. 2001. Transnational migration: taking stock and future directions. Global networks 1(3):195-216.

Levitt P, Nyberg-Sorensen N. 2004. The transnational turn in migration studies. Global Migration Perspectives. No. 6. Global Commission on International Migration. www.transnational-studies.org/pdfs/global_migration_persp.pdf

López E, Bocco G, Mendoza M, Velázquez A, Aguirre-Rivera JR. 2006. Peasant emigration and land-use change at the watershed level: A GIS-based approach in Central Mexico. Agricultural Systems 90(1-3):62-78.

Lowell BL, de la Garza R. 2000. The development role of remittances in US Latino communities and in Latin American countries. Final project report to the Inter-American Dialogue. Georgetown, Texas: Institute for the Study of International Migration, Georgetown University Tomás Rivera Policy Institute, Department of Government, University of Texas.

Martin PL, Taylor JE. 1996. The anatomy of the migration hump. In: Taylor JE, editor. Development strategy, employment, and migration: insights from models. Paris: OECD Development Centre. p 43-62.

Martinez C. 2004. Interview with Lic. Cesar Martinez. Head of the Procuraduria Agraria in Autlán. Autlán (24 March 2004).

Masera OR, Ordoñez MJ, Dirzo R. 1997. Carbon emissions from Mexican forests: Current situation and long-term scenarios. Climatic Change 35:265-295.

Massey DS. 1985. The settlement process among Mexican migrants to the United States: new methods and findings. In: Levine DB, Warren R, editors. Immigration statistics: A story of neglect. Washington DC: National Academy Press. p 255-292.

Massey DS, Arango J, Hugo G, Kouaouci A, Pellegrino A, Taylor JE. 1998. Worlds in Motion. Understanding international migration at the end of the millennium. Oxford: Clarendon Press. 362 p.

Massey DS, Basem LC. 1992. Determinants of savings, remittances, and spending patterns among US migrants in four Mexican communities. Sociological Inquiry 62(2):185-207.

Massey DS, Durand J, Malone NJ. 2002. Beyond smoke and mirrors: Mexican immigration. New York: Sage Foundation.

Massey DS, España FG. 1987. The social process of international migration. Science 2(4816):733-738.

Massey DS, Parrado EA. 1998. International migration and business formation in Mexico. Social Science Quarterly 79(1):1-20.

Mather AS, Needle CL, Fairbairn J. 1999. Environmental kuznets curves and forest trends. Geography 84:55-65.

McDowell C, de Haan A. 1997. Migration and Sustainable Livelihoods: A Critical Review of the Literature. Brighton: Institute for Development Studies. Report nr 65. 29 p.

McKay D. 2003. Cultivating New Local Futures: Remittance Economies and Land-use Patterns in Ifugao, Philippines. Journal of Southeast Asian Studies 34(2):285-306.

McKay D. 2005. Reading remittance landscapes: Female migration and agricultural transition in the Philippines. Danish Journal of Geography 105(1):89-99.

Meadows DH. 1972. Limits to growth. New York: Universe Books.
Medina E. 2004. Interview with Ernesto Medina. Historian of Autlán. Autlán (15 April 2004).
Mejia J. 2003. Interview with Justino Mejia. President of the municipality of Tonaya. Tonaya (6 February 2003).
Mertz O, Wadley RL, Christensen AE. 2005. Local land use strategies in a globalizing world: Subsistence farming, cash crops and income diversification. Agricultural Systems 85(3):209-215.
Messerli P, Wiesmann U. 2004. Synopsis of Syndrome Contexts and Core Problems Associated with Syndromes of Global Change. In: Hurni H, Wiesmann U, Schertenleib R, editors. Research for Mitigating Syndromes of Global Change. A Transdisciplinary Appraisal of Selected Regions of the World to Prepare Development-Oriented Research Partnerships. Berne: Geographica Bernensia. p 383-423.
Meyer WB, Turner BL. 1994. Changes in Land Use and Land Cover: A Global Perspective. Cambridge: Cambridge University Press.
Millennium Ecosystem Assessment. 2005. Ecosystems and Human Well-being: Synthesis. Washington, DC: Island Press. 155 p.
Millington AC, Velez-Liendo XM, Bradley AV. 2003. Scale dependence in multitemporal mapping of forest fragmentation in Bolivia: implications for explaining temporal trends in landscape ecology and applications to biodiversity conservation. Journal of Photogrammetry and Remote Sensing 57:289-299.
Mines R, de Janvry A. 1982. Migration to the United States and Mexican rural development: A case study. American Journal of Agricultural Economics 64(3):444-454.
Mohr GM. 2002. Blue agave and its importance in the tequila industry. Ethnobotanical Leaflets. www.siu.edu/~ebl/leaflets/agave.htm
Moran EF, Siqueira A, Brondizio E. 2003. Household demographic structure and its relationship to deforestation in the Amazon Basin. In: Fox J, Rindfuss RR, Walsh SJ, Mishra V, editors. People and the Environment: Approaches for linking household and community surveys to remote sensing and GIS. Dordrecht, Netherlands: Kluwer Academic Publishers. p 61-89.
Mosse D, Gupta S, Mehta M, Shah V, Rees J. 2002. Brokered livelihoods: debt, labour migration and development in tribal western India. The Journal of Development Studies 38(5):59-88.
Municipality of Autlán. 2001. Guía Turistica del Municipio de Autlán. Adminstracion 2001-2003. Presidencia municipal. Regiduría de Turismo, Dirección de Promoción, Direccíon de Comunicación.
Municipality of Autlán. 2002. Plan de desarrollo del municipio de Autlán de Navarro, Jalisco. Autlán: Municipality of Autlán. 401 p.
Municipality of Autlán. 2003. Proyecciones de población. Cadastre.

Murray-Li T. 2002. Engaging simplifications: community-based resource management, market processes and state agendas in upland Southeast Asia. World Development 30(2):265-283.

Nadal A. 2002. The Environmental and Social Impacts of Economic Liberalization on Corn Production in Mexico. Gland, Switzerland: Oxfam GB and WWF International.

Nichols S. 2002. "Another Kind of Remittance: Transfer of Agricultural Innovations by Migrants to Their Communities of Origin". Paper presented at the Second Colloquium on International Migration: Mexico-California. University of California, Berkeley (29 March).

Nicholson B. 2004. Migrants as Agents of Development: Albanian Return Migrants and Micro-enterprise. In: Pop D, editor. New Patterns of Labour Migration in Central and Eastern Europe. Cluj Napoca, Romania: AMM Editura. p 94-110.

Nii Addy D, Wijkström B, Thouez C. 2004. Migrant Remittances: Country of Origin Experiences - Strategies, Policies, Challenges and Concerns. The International Migration Policy Programme. www.livelihoods.org

Nuijten M. 2001. What's in the land? The multiple meanings of land in a transnationalized Mexican village. In: Zoomers A, editor. Land and sustainable livelihoods in Latin America: Koninklijk Instituut Voor de Tropen Vervuert Verlag. p 71-93.

Nüsser M. 2003. Naturraum, Ressourcennutzung und Umweltdegradation: Mensch-Umwelt-Beziehungen in subtropischen Hochgebirgsregionen (Nord-Pakistan und Lesotho). Zürich: University of Zürich (unpublished). 33 p.

Nyberg-Sørensen N, Hear NV, Engberg-Pedersen P. 2002. The Migration-Development Nexus. Evidence and Policy Options. State-of-the-Art Overview. International Migration 40(5):3-43.

Olesen H. 2002. Migration, Return, and Development: An Institutional Perspective. International Migration 40(5):125-150.

Orozco M. 2002. Worker Remittances: the human face of globalization. Working paper commissioned by the Multilateral Investment Fund of the Inter-American Development Bank, October 10, 2002. www.thedialogue.org/publications/country_studies/remittances/worker_remit.pdf

Orozco M. 2003. Costs, economic identity and banking the unbanked. Testimony presented before the Congressional Hispanic Caucuses March 26, 2003. Inter American Dialogue.

Pan D, Domon G, de Bois S, Bouchard A. 1990. Temporal (1958-1993) and spatial patterns of land use changes in Haut Saint-Laurent (Quebec, Canada) and their relation to landscape physical attributes. Landscape Ecology 14:35-52.

Pan WKY, Bilsborrow RE. 2005. The use of a multilevel statistical model to analyze factors influencing land use: a study of the Ecuadorian Amazon. Global and Planetary Change 47:232-252.

Peets R, Watts M. 1996. Liberation ecologies - environment, development, social movements. London: Routledge.

Petschel-Held G, Block A, Schellnhuber H-J. 1995. Syndrome des Globalen Wandels - Ein systemarer Ansatz für Sustainable-Development-Indikatoren. Geowissenschaften 13(3):81-87.
Portes A, Sensebrenner J. 1993. Embeddedness and Immigration: Notes on the Social Determinants of Economic Action. American Journal of Sociology 98(6):1320-1350.
Portes A, Guranizo LE, Landolt P. 1999. The study of transnationalism: pitfalls and promise of an emergent research field. Ethnic and Racial Studies 22(2):217-237.
Portner B. 2005. Land use strategies of migrant and non-migrant households in western Mexico [MSc thesis]. Berne: University of Berne (unpublished). 159 p.
Proctor JD. 1998. The meaning of global environmental change: retheorizing culture in human dimensions research. Global Environmental Change 8:227-248.
Procuraduria Agraria. 1995. Glosario de Términos Jurídicos. México: INEGI. www.inegi.gob.mx.
Pushpangadan K. 2003. Remittances, consumption and economic growth in Kerala: 1980-2000. Thiruvananthapuram: Centre for Development Studies. 31 p.
Ratha D. 2004. Enhancing the developmental effect of workers' remittances to developing countries. Global development finance. Washington, DC: World Bank. p 169-173.
Riak-Akuei S. 2005. Remittances as unforeseen burdens: the livelihoods and social obligations of Sudanese refugees. Geneva: Global Commission on International Migration. 16 p.
Rindfuss PR, Stern PC. 1998. Linking remote sensing and social science: the need and the challenges. In: Liverman D, Moran EF, Rindfuss RR, Stern PC, editors. Linking remote sensing and social science. Washington, DC: National Academy Press. p 1-27.
Rockwell RC. 1994. Culture and cultural change. In: Myer WB, Turner BL, editors. Changes in land use and land cover: a global perspective. Cambridge, UK: Cambridge University Press. p 357-382.
Rogaly B, Coppard D, Safique A, Rana K, Sengupta A, Biswas J. 2002. Seasonal Migration and Welfare/Illfare in Eastern India: A Social Analysis. The Journal of Development Studies 38(5):89 - 114.
Rudel TK, Coomes OT, Moran E, Achard F, Angelsen A, Xu J, Lambin EF. 2005. Forest transitions: Towards a global understanding of land use change. Global Environmental Change 15:23-31.
Sadoulet E, de Janvry A, Davis B. 2001. Cash transfer programs with income multipliers: PROCAMPO in Mexico. FCND Discussion paper no. 99: IFPRI. www.ifpri.org/divs/fcnd/dp/papers/fcndp99.pdf
SAGARPA. 2002. PROCAMPO: Vamos al grano para progressar. Mexican Secretary of Agriculture, Ranching, Rural Development, Fisheries and Food Supply (SAGARPA). www.procampo.gob.mx
SAGARPA. 2004. Statistics on agriculture, cattle ranching, rural development, fisheries. Mexican Secretary of Agriculture, Ranching, Rural Development, Fisheries and Food Supply (SAGARPA). www.sagarpa.gob.mx

SAGARPA. 2006. Agricultural statistics established by Lic. Salvador Hernández Pelayo. Coordinador de Informática y Estadística. Distrito de Riego 05. SAGARPA. Mexican Secretary of Agriculture, Ranching, Rural Development, Fisheries and Food Supply (SAGARPA).

Sánchez Reaza J, Rodríguez Pose A. 2002. The Impact of Trade Liberalization on Regional Disparities in Mexico. Growth and Change 33(1):72-90.

Sander C. 2003. Migrant remittances to developing countries. A scoping study: Overview and introduction to issues for pro-poor financial services. DFID. www.livelihoods.org

Sander C, Munzele SM. 2004. Migrant Labor Remittances in Africa: Reducing Obstacles to Developmental Contributions. The World Bank Africa Region Working Paper Series. www.worldbank.org

SARPI. 2000. NCCR North-South: Research partnerships for mitigating syndromes of global change. Proposal submitted to the Swiss National Science Foundation for a National Centre of Competence in Research (NCCR). Berne: Swiss Association of Research Partnership Institutions (SARPI).

Sayer A. 1984. Methods in Social Science. A Realist Approach. London: Hutchinson.

Schubert J. 2005. Political ecology in development research. An introductory overview and annotated bibliography. Bern:NCCR North-South. 66 p.

Scoones I. 1998. Sustainable rural livelihoods. A framework for analysis. IDS Working paper(72):20.

Scoones I. 1999. New Ecology and the Social Sciences: What Prospects for a Fruitful Engagement? 28:479-507.

Seddon D. 2004. South Asian remittances: implications for development. Contemporary South Asia 13(4):403-420.

SEMADES. 2006. La deforestacion en Jalisco. SEMADES (*Secretaría de Medio Ambiente para el Desarrollo Sustentable*; The Department of the Environment for Sustainable Development). Guadalajara.
http://semades.jalisco.gob.mx/site/moet/index.htm

SEMARNAT. 2002. Dimension ambiental. SEMARNAT (*Secretaría de Medio Ambiente y Recursos Naturales*; The Secretariat of Environment and Natural Resources). Subsecretaría de Gestión para la Protección Ambiental, Dirección General de Federalización y Descentralización de Servicios Forestales y de Suelo. www.semarnat.gob.mx

Serneels S, Lambin EF. 2001. Proximate causes of land-use change in Narok District, Kenya: a spatial statistical model. Agriculture, Ecosystems and Environment 85:65-81.

Silbernagel J, Martin SR, Gale MR, Chen J. 1997. Prehistoric, historic, and present settlement patterns related to ecological hierarchy in the Eastern Upper Peninsula of Michigan, USA. Landscape Ecology 12:223-240.

Simonian L. 1995. Defending the land of the jaguar: a history of conservation in Mexico. Austin: The University of Texas Press. 326 p.

Snyder R. 1998. The future role of the ejido in rural Mexico. La Jolla: Center for U.S. Mexican Studies.

Soini E. 2005. Land use change patterns and livelihood dynamics on the slopes of Mt. Kilimanjaro, Tanzania. Agricultural Systems 85(3):306-323.

Soliva R. 2000. Der Naturschutz in Nepal. Eine akteurorientierte Untersuchung aus der Sicht der Politischen Oekologie. Kultur, Gesellschaft, Umwelt. Schriften zur Südasien- und Südostasien-Forschung 5. University of Zürich.

Sorensen NN, Olwig KF. 2002. Work and Migration: Life and livelihoods in a globalizing world. London: Routledge.

Southworth J, Tucker C. 2001. The influence of accessibility, local institutions, and socioeconomic factors on forest cover change in the mountains of western Honduras. Mountain Research and Development 21(3):276-283.

Stéphenne N, Lambin E. 2001. A dynamic simulation model of land-use changes in Sudano-sahelian countries of Africa (SALU). Agriculture, Ecosystems and Environment 85:145-161.

Stéphenne N, Lambin EF. 2004. Scenarios of land-use change in Sudano-sahelian countries of Africa to better understand driving forces. GeoJournal 61(4):365-379.

Stromph TJ, Fresco LO, van Keulen H. 1994. Land use systems evaluation: concepts and methodology. Agricultural Systems 44:243-255.

Taylor EJ. 1999. The New Economics of Labour Migration and the Role of Remittances in the Migration Process. International Migration 37(1):63-88.

The Economist. 2002 The longest journey: A survey of migration. The Economist (2 November 2002):1-16.

Thieme S. 2006. Social Networks and Migration: Far West Nepalese Labour Migrants in Delhi. Münster, London: LIT Publishing house.

Thieme S, Müller-Böker U. 2004. Financial self-help associations among Far West Nepalese labor migrants in Delhi, India. Asian and Pacific Migration Journal 13(3):339-361.

Thieme S, Wyss S. 2005. Migration Patterns and Remittance Transfer in Nepal: A Case Study of Sainik Basti in Western Nepal. International Migration 43(5):59-98.

Thomas-Hope E. 2002. Skilled labour migration from developing countries: Study on the Caribbean Region. International Migration Paper No. 50. Geneva: International Labour Office.

Tiemoko R. 2004. Migration, Return and Socio-Economic Change in West Africa: The Role of the Family. The Sussex Centre for Migration Working Paper. www.sussex.ac.uk

Tienda M, Verduzco G, Greenwood M, Unger K. 1997. Mexico and migration effects. Migration Between Mexico and the United States: Binational Study. Mexico City and Washington: Mexican Ministry of Foreign Affairs and U.S. Commission on Immigration Reform.

Tiffen M, Mortimer M, Gichuki F. 1994. More people, less erosion. New York: Wiley.

Torres G. 1992. Plunging into the garlic. Methodological issues and challenges. In: Long N, Long A, editors. Battlefields of knowledge. The interlocking of theory and

practice in social research and development. London and New York: Routledge Press. p 85-114.
Trigueros P, Rodriguez J. 1988. Migración y vida familiar en Michoacán. In: Lopez G, editor. Migracion en el occidente de México. Zamora: Colegio de Michoacán. p 201-221.
Turner BL, R. H. Moss, and D. L. Skole. 1993. Relating land use and global land-cover change: A proposal for an IGBP-HDP core project. Stockholm: Joint publication of the International Geosphere-Biosphere Programme (Report No. 24) and the Human Dimensions of Global Environmental Change Programme (Report No. 5).
Turner BL. 1997. The sustainability principle in global agendas: Implications for understanding land-use/cover change. The Geographical Journal 163(2 Environmental Transformations in Developing Countries):133-140.
Turner BL. 2001. Towards Integrated Land-Change Science: Advances in 1.5 decades of sustained international research on land-use and land-cover change. In: Steffen W, editor. Advances in global environmental change research. Berlin, New York: Springer Verlag.
Turner BL, Skole D, Fischer G, Leemans R. 1995. Land-Use and Land-Cover Change; Science/Research Plan. Stockholm and Geneva: IGBP and IHDP.
Turner MG, Wear DN, Flamm RO. 1996. Land ownership and land-cover change in the southern Appalachian highlands and the Olympic Peninsula. Ecological Applications 6(4):1150-1172.
UNEP. 1999. Global Environment Outlook 2000. New York: United Nations Environment Programme, Oxford University Press. 398 p.
Vargas Martín E. 2003. Interview with the Lic. Enrique Vargas Martín. Local manager for the company "*Azul, Agricultura and Servicios*". Autlán (18 February 2003).
Vasquez GJA, Cuevas GR, Cochrane TS, Iltis HH, Santana MFJ, Guzmán HL. 1995. Flora de Manantlán. Plantas vasculares de la Reserva de la Biosfera Sierra de Manantlán. Texas, USA: Botanical Research Institute of Texas.
Vayda A, Walters B. 1999. Against Political Ecology. Human Ecology 27(1):167-179.
Veldkamp A, Lambin EF. 2001. Editorial: Predicting land use change. Agriculture, Ecosystems and Environment 85(1-3):1-6.
Verbist B, Putra AED, Budidarsono S. 2005. Factors driving land use change: Effects on watershed functions in a coffee agroforestry system in Lampung, Sumatra. Agricultural Systems 85(3):254-270.
Verburg PH, Veldkamp A. 2005. Introduction to the Special Issue on Spatial modeling to explore land use dynamics. International Journal of Geographical Information Science 19(2):99-102.
Verburg PH, Veldkamp A, Fresco LO. 1999. Simulation of changes in the spatial pattern of land use in China. Applied Geography 19:211-233.
Verduzco G, Unger K. 1998. Impacts of Migration in Mexico. Migration Between Mexico and the United States: Binational Study. Mexico City and Washington, D.C.: Mexican Ministry of Foreign Affairs and U.S. Commission on Immigration Reform. p 395-436.

Vertovec S. 1999. Conceiving and researching transnationalism. Ethnic and Racial Studies 22:447-462.

Vidriales Guzmán M. 2003. Interview with María Vidriales Guzmán. *Ejidataria* in El Jalocote. El Jalocote (4 March 2003).

Vitousek PM, Mooney HA, Lubchenco J, Melillo JM. 1997. Human domination of earth's ecosystems. Science 277(25):494-499.

Walker PA. 2005. Political ecology: where is the ecology? Progress in Human Geography 29(1):73-82.

Walker R, Perz S, Caldas M, Teixeira Silva LG. 2002. Land use and land cover change in forest frontiers: The role of household life cycles. International Regional Science Review 25(2):169-199.

Walsh SJ, Evans TP, Welsh W, Entwisle B, Rindfuss RR. 1999. Scale-dependent relationships between population and environment in northeast Thailand. Photogrammetric Engineering and Remote Sensing 51(1):97-105.

Wanner P. 2006. Transferts de fonds: migration Nord-Sud, outil pour la réduction de la pauvreté. Le temps. 18 May 2006;20.

WBGU. 1997. World in Transition: The Research Challenge. Berlin: Springer Verlag.

Weischet W, Caviedes CN. 1993. The persisting ecological constraints of tropical agriculture. Essex: Longman Group.

Woodwell GM, Hobbie JE, Houghton RA, Melillo JM, Moore B. 1983. Global deforestation: contribution to atmospheric carbon dioxide. Science 222:1081-1086.

World Bank. 2004. Global Development Finance. Washington, DC: World Bank.

World Resources Institute. 2000. World resources 2000-2001: People and ecosystems: The fraying web of life. Washington: World Resources Institute. 389 p.

UNEP. 1999. Global Environment Outlook 2000. New York: United Nations Environment Programme, Oxford University Press. 398 p.

Young M. 2002. Land use change in the context of a cattle breeding revolution. Transnational migration and a Biosphere Reserve [MSc. thesis]. Wisconsin: University of Wisconsin (unpublished). 160 p.

Zoomers A. 2001. Land and Sustainable Livelihood in Latin America. Koninklijk Instituut Voor de Tropen. Vervuert Verlag. 257 p.

Definitions / Glossary

Agave azul: *Agave tequilana* Weber (Agavaceae).

Caciques/caciquismo: Caciquismo is used to refer to a dominant relation with a local leader, landowner or local politician (the so-called cacique). It conveys the idea of a degree of economic or political power, but there is a strong implication of 'influence' and the capacity to manipulate other people's actions (Torres 1992).

Cerro: The hills (Gerritsen 2002).

Coamil: Maize cultivation through shifting cultivation practices (Gerritsen 2002).

Comunidad agraria: *Comunidades agrarias* [agrarian communities] are collective owners of their land under a common property regime. It is the population center constituted by the land, forests and water sources, that were recognized or restituted to the agrarian community, and of which the community presumably had possession of over centuries. (Procuraduria Agraria 1995).

Coyote: Term for people who help migrants to illegally cross the border between Mexico and the United States for payment.

Driving forces: Influences on land use change are divided into proximate causes and underlying driving forces (Geist and Lambin 2001). Proximate causes are human activities (land use) that directly affect the environment and thus cause land use change, such as an expansion of agriculture. Underlying driving forces are fundamental forces that underpin the proximate causes of land use and land cover change. These underlying driving forces include socio-economic, political and biophysical processes that directly or indirectly affect the decision-making of the land user (adapted from Hansen 2005).

Ejido: An *ejido* is a communal form of land tenure to which members have use rights, usually in the form of individual plots of land. The formation of *ejidos* since the Mexican Revolution of 1910 has involved the transfer of over 70 million hectares from large estates (*haciendas*) to approximately three million beneficiaries (Snyder 1998).

Ejidatarios/ejidatarias: Members of the *ejido* with land rights.

El Norte: The North. Term used by Mexicans to refer to the USA.

Global change: Global-scale human, human-induced and natural changes that modify the functionality of the natural, social, economic and cultural dimensions of the Earth system (Hurni et al. 2004).

Globalization: The growing and accelerated interconnectedness of the world in an economic, political, social and cultural sense (Global Land Project 2005).

Hacienda: Large agricultural estates in Mexico that existed until the Revolution in 1910.

Household: Refers to a nuclear household. Households with at least one immediate family member in the United States were classified as international migrant households. A migrant household is when the member in migration would live with the household if not in migration.

International labor migrant: A person who is to be engaged, is engaged or has been engaged in a remunerated activity in a State of which he or she is not a national (UN-Convention on the Protection of the Rights of All Migrant Workers and their Families, Article 2, 1990, in: Thieme 2005).

Land cover: The physical characteristics of earth's surface, captured in the distribution of vegetation, water, desert, ice, and other physical features of the land, including those created solely by human activities such as mine exposure and settlement (Lambin et al. 1999).

Land cover changes: Changes in the biological and physical cover of the earth's surface in terms of vegetation or man-made features. Land cover changes can be divided into land cover modifications and land cover conversions. According to Lambin et al. (2003) "Land cover conversions (i.e. the complete replacement of one cover type by another) are measured by a shift from one land cover category to another, as is the case in agricultural expansion, deforestation, or changes in urban cover. Land cover modifications are more subtle changes that affect the character of the land cover without changing its overall classification" (adapted from Hanson 2005). For instance, the change from subsistence crop to cash crop cultivation systems constitutes a land cover modification within the land cover class of agriculture. In this thesis the term *land use change* is used to describe *land cover modifications* and *land cover conversions*.

Land degradation: From an agricultural perspective, land degradation is defined as a reduction in the soil's capacity to produce crops or biomass for livestock. From an ecological perspective, land degradation is damage to the healthy functioning of land-based ecosystems.

Land use: The intended utilization and management strategy of human agents, or land managers, in regards to land cover type. Shifts in management constitute land use changes (Lambin et al. 1999).

Migrant (short-term): Persons who move to a country other than that of his or her usual residence for a period of at least three months but less than a year (Alfieri and Havinga 2005).

Migrant (long-term): Persons who move to a country other than that of his or her usual residence for a period of at least a year, so that the country of destination effectively becomes his or her new country of usual residence (Alfieri and Havinga 2005).

Migration (circular): Circular migration is short-term (1–6 months), often seasonal labor migration that does not involve a permanent change in residence (Chapman and Prothero 1985).

New Economics of Labor Migration (NELM): New economics of labor migration (NELM) theorists propose that rural communities located in areas with high quality land, better infrastructure, and good access to markets offer enhanced opportunities for profitable investment and therefore are more likely to stimulate migration (Basok 2003).

Parcela: Land plot (Appendini 2001).

Positivism: A philosophy of science originally proposed by August Comte in the early 19th century. Its primary purpose was to distinguish science from metaphysics and religion. Broadly, it accepts that: (a) scientific statements should be based on empirical observations and facts; (b) the (mostly quantitative) methods of the natural sciences can be extended to the study of social phenomena; (c) general, universal laws are the ultimate goal of scientific inquiry, i.e. the search for empirical regularities, for law and order (Briassoullis 1999).

Realism: A philosophy of science which uses abstraction to identify the necessary/causal powers of specific structures which are realized under contingent/specific conditions. It regards the world as differentiated, stratified, and made up not only of events (as positivism does) but also of mechanisms and structures. Structures are seen as sets of inter-related objects that have essential properties and thus characteristic ways of acting. That is they possess "causal powers and liabilities" (Sayer 1984) by virtue of what they are and which are, thus, necessary. Realist analysis tries to identify causal chains which place particular events within these deeper mechanisms and structures (Briassoulis 1999).

Remittances (Individual): Transfers of assets by members of immigrant communities or foreign nationals from the country where they live and work back to relatives or other individuals in their country of origin (adapted from Seddon 2004).

Remittances (Collective): Transfers of assets by a group of migrants to their community of origin to support community and development projects or to contribute to disaster-related recovery efforts (Lowell and de la Garza 2000).

Reverse leasing arrangements: Smallholders rent their land plots to independent contractors (often affiliated or directly owned by large transnational corporations) who bring in capital, machinery, labor, and other inputs needed for agricultural production. The independent contractor, and not the landowner, takes control of all management decisions (Bowen 2004).

Annex

Annex 1

Cross-tabulation matrix of land use changes 1990-2000 in the municipality of Autlán (ha)

2000 -> 1990	Agriculture	Bare ground	Dry forest	Pastures	Pine-oak forest	Urban	<-2000 TOTAL
Agriculture	11952	0	259	731		29	12972
Bare ground	2	23	148	106	104		382
Dry forest	226	68	20898	2388	25	19	23625
Pastures	1006	69	5874	4530	46	70	11596
Pine-oak forest		6	2	1	23984		23993
Urban	48		9	56	0	991	1104
TOTAL	13234	167	27190	7813	24159	1110	73673

Legend: Land use transition matrix between 1990 and 2000. Each cell is read as a transition from the horizontal category (1990) to the vertical category (2000). The numbers present the proportion of a particular land use (in ha) from 1990 which changed or remained stable over the study period.

Source: Automatic calculations by software program *Manifold 6.5* based on computation of differences between digitized satellite images of 1990 and 2000.

Die VDM Verlagsservicegesellschaft sucht für wissenschaftliche Verlage abgeschlossene und herausragende

Dissertationen, Habilitationen, Diplomarbeiten, Master Theses, Magisterarbeiten usw.

für die kostenlose Publikation als Fachbuch.

Sie verfügen über eine Arbeit, die hohen inhaltlichen und formalen Ansprüchen genügt, und haben Interesse an einer honorarvergüteten Publikation?

Dann senden Sie bitte erste Informationen über sich und Ihre Arbeit per Email an *info@vdm-vsg.de*.

Sie erhalten kurzfristig unser Feedback!

VDM Verlagsservicegesellschaft mbH
Dudweiler Landstr. 99
D - 66123 Saarbrücken

Telefon +49 681 3720 174
Fax   +49 681 3720 1749

www.vdm-vsg.de

Die VDM Verlagsservicegesellschaft mbH vertritt

Printed by Books on Demand GmbH, Norderstedt / Germany